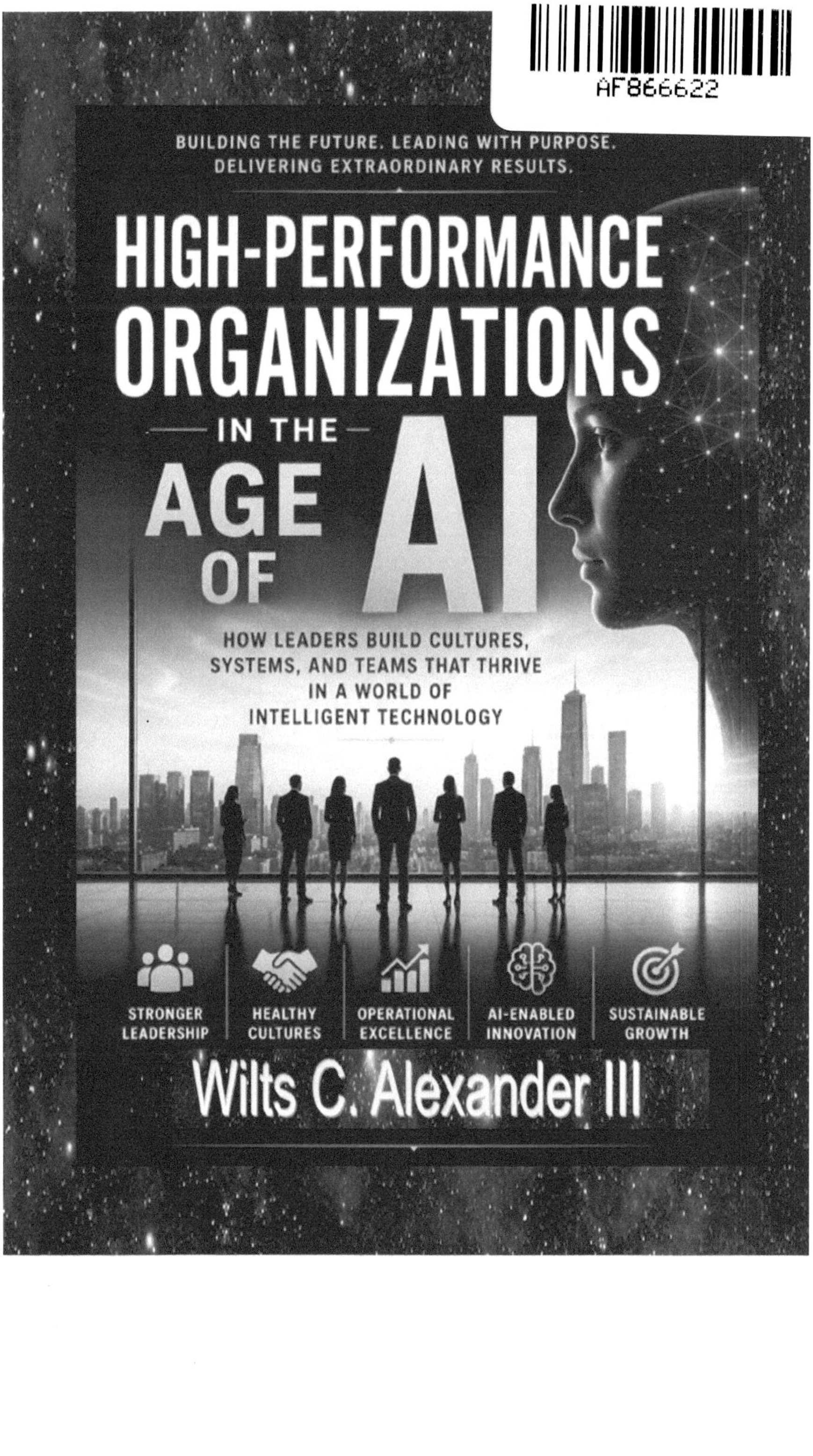
BUILDING THE FUTURE. LEADING WITH PURPOSE.
DELIVERING EXTRAORDINARY RESULTS.
HIGH-PERFORMANCE
ORGANIZATIONS
IN THE
AGE OF AI
HOW LEADERS BUILD CULTURES, SYSTEMS, AND TEAMS THAT THRIVE IN A WORLD OF INTELLIGENT TECHNOLOGY
STRONGER LEADERSHIP
HEALTHY CULTURES
OPERATIONAL EXCELLENCE
AI-ENABLED INNOVATION
SUSTAINABLE GROWTH
Wilts C. Alexander III

Published by RICHER Press

An Imprint of Richer Life, LLC
5710 Ogeechee Road, Suite 200-175, Savannah, Georgia 31405
www.richerlifellc.com

Cover Design: RICHER Media

Volume book discounts are available for groups, companies and organizations. Contact the publisher for information and order instructions.

Library of Congress Control Number: 2026947620

High-Performance Organizations in The Age of AI
How Leaders Build Thriving Cultures, Systems and Teams

Wilts C. Alexander III

1. Business Decision Making 2. Decision-Making & Problem Solving
3. Business Ethics

(pbk : alk. Paper)

ISBN: 979-8-9928996-8-9 Paperback

PRINTED IN THE UNITED STATES OF AMERICA

Dedication

To the leaders who refuse to be left behind.

To the executives, entrepreneurs, managers, educators, public servants, and change agents who understand that the future does not belong to those who wait, it belongs to those who learn, adapt, and lead.

May this book help you build organizations that create greater value, develop extraordinary people, and use technology to improve the human condition.

And to the leaders who came before us, those who taught us that character, service, accountability, and vision matter more than any tool ever invented, thank you for laying the foundation upon which we build.

TABLE OF CONTENTS

INTRODUCTION

The New Competitive Advantage

Why This Book Matters for Leaders
Organizations, and the Future of work

Every few decades, a technological shift emerges that fundamentally changes the way organizations operate, compete, and create value.

The steam engine transformed manufacturing.

Electricity transformed industry.

The internet transformed communication and commerce.

Today, artificial intelligence is transforming everything.

The headlines often focus on the technology itself, new models, new capabilities, and new applications appearing almost daily. Leaders are bombarded with predictions ranging from limitless opportunity to existential risk. Some are excited. Some are overwhelmed. Many are unsure where to begin.

Yet amid all the noise, one truth is becoming increasingly clear:

The organizations that thrive in the age of AI will not necessarily be those with the best technology. They will be the organizations with the best leadership, the most adaptable cultures, the clearest strategies, and the strongest ability to integrate AI into how people work, learn, and create value.

This is an important distinction.

Many leaders mistakenly view AI as a technology project. They delegate it to IT departments, technology teams, or outside consultants and hope innovation somehow emerges from the effort.

But AI is not merely a technology initiative.

It is an organizational transformation initiative.

The introduction of artificial intelligence affects decision-making, communication, operations, customer experience, workforce development, organizational structure, leadership practices, and competitive strategy. It influences how work gets done, who does it, and what skills become most valuable.

In other words, AI changes the organization itself.

That reality presents both a challenge and an opportunity.

The challenge is that many organizations were designed for a different era. Their structures, processes, management systems, and leadership approaches were built for a world where information moved more slowly, decisions were made with less data, and innovation cycles took years instead of months.

The opportunity is that organizations willing to adapt can achieve levels of performance previously unimaginable.

Teams can become more productive.

Decision-making can become faster and more informed.

Customer experiences can become more personalized.

Innovation can accelerate.

Resources can be allocated more intelligently.

Entire business models can be reinvented.

Yet none of these benefits occur automatically.

Technology alone does not create transformation.

People do.

History repeatedly teaches us that new tools only create value when leaders help people adopt them, trust them, and integrate them into meaningful work. Organizations that focus exclusively on technology often fail because they neglect the human systems that determine whether change succeeds or fails.

This book was written to address that challenge.

It is not intended to be a technical manual.

You will not find complex programming instructions, detailed algorithms, or advanced computer science concepts in these pages.

Instead, this book focuses on a more important question:

How do leaders design and build high-performance organizations that successfully leverage artificial intelligence while strengthening the human capabilities that drive long-term success?

Whether you are a CEO, entrepreneur, executive, manager, board member, consultant, or organizational leader, the principles in this book are designed to help you think strategically about the future of your organization.

You will learn:

- Why AI represents an organizational transformation rather than a technology project.
- The characteristics of high-performance organizations in the AI era.
- How culture influences successful AI adoption.
- The leadership skills required to guide transformation.
- How to create trust during periods of rapid change.
- Ways to use AI to improve productivity, innovation, and decision-making.

- How to prepare your workforce for an AI-enabled future.
- Practical steps for beginning your organization's AI journey.

Most importantly, you will discover that the future belongs neither to humans nor to machines alone.

It belongs to organizations that learn how to combine the strengths of both.

The most successful companies of the next decade will not replace human intelligence with artificial intelligence. They will augment human intelligence with artificial intelligence.

They will create environments where technology amplifies creativity, improves decision-making, accelerates learning, and frees people to focus on the highest-value activities.

They will build cultures that embrace innovation while preserving trust.

They will move quickly without losing their values.

They will become learning organizations capable of adapting continuously as technology evolves.

That is the central thesis of this book.

The future will not be won by organizations that merely adopt AI.

The future will be won by organizations that redesign themselves to thrive alongside it.

The good news is that this transformation is accessible to organizations of every size.

You do not need to be a global technology company.

You do not need a billion-dollar budget.

You do not need an army of data scientists.

What you do need is leadership, clarity, commitment, and a willingness to learn.

The organizations that start today will gain a significant advantage tomorrow.

Those who delay may find themselves struggling to catch up.

The age of AI is not coming.

It is already here.

The question is not whether artificial intelligence will change your organization.

The question is whether you will intentionally lead that change, or be forced to react to it.

This book is designed to help you lead.

IN THE
AGE
OF
AI

PART I
UNDERSTANDING THE AI REVOLUTION

IN THE
AGE
OF
AI

CHAPTER 1

The Age of AI Has Arrived

What Every Leader Needs to Understand

A few years ago, artificial intelligence was largely a topic discussed by technology companies, researchers, and futurists.

Today, it is a boardroom conversation.

CEOs are discussing it.

Investors are evaluating it.

Employees are experimenting with it.

Customers are beginning to expect it.

Competitors are implementing it.

Whether leaders feel ready or not, artificial intelligence has moved from the laboratory into everyday business operations.

The question is no longer whether AI will impact your organization.

The question is how much.

For many leaders, this realization creates both excitement and uncertainty.

They hear stories about organizations dramatically increasing productivity, reducing costs, accelerating innovation, and creating entirely new products and services. At the same time, they hear warnings about job displacement, cybersecurity risks, ethical concerns, and the possibility of being left behind.

Both perspectives contain some truth.

Like every major technological advancement before it, artificial intelligence is neither inherently good nor inherently bad. It is a tool.

And like any tool, its value depends on how effectively it is used.

The challenge for leaders is determining what AI actually means for their organization and how to respond wisely.

The Pattern We Have Seen Before

Throughout history, transformational technologies have followed a predictable pattern.

At first, they are dismissed.

Then they are misunderstood.

Then they are overhyped.

Eventually, they become integrated into everyday life.

When electricity was first introduced, many organizations struggled to understand how it would impact their operations.

When computers entered the workplace, some executives viewed them as expensive calculators.

When the internet emerged, many believed it would remain a niche technology with limited business applications.

In each case, the organizations that learned fastest gained a significant advantage.

The organizations that resisted change often found themselves struggling to remain competitive.

Artificial intelligence appears to be following the same pattern.

Some organizations are already integrating AI into customer service, marketing, operations, finance, product development, human resources, and strategic planning.

Others remain on the sidelines, unsure whether AI represents a genuine opportunity or simply another technology trend.

History suggests that waiting too long can be costly.

The leaders who succeed are rarely those who predict the future perfectly.

They are the ones who prepare for multiple futures and adapt quickly as events unfold.

What AI Really Is

One reason AI generates confusion is that people often use the term to describe many different technologies.

For the purposes of this book, artificial intelligence can be thought of as systems that perform tasks traditionally requiring human intelligence.

These tasks include:

- Analyzing information
- Identifying patterns
- Making predictions
- Generating content
- Answering questions
- Supporting decisions
- Automating routine activities
- Learning from large amounts of data

AI can draft reports.

Summarize meetings.

Analyze contracts.

Forecast demand.

Recommend actions.

Generate marketing content.

Assist software developers.

Support customer service representatives.

Help managers evaluate information more quickly.

What AI cannot do particularly well is replace human judgment, wisdom, leadership, empathy, creativity, values, ethics, trust-building, and relationship management.

Those capabilities remain deeply human.

The organizations that understand this distinction will make better decisions than those who assume AI can solve every problem.

Why This Time Is Different

Many technology innovations improve a specific function.

AI is different because it touches nearly every function.

Marketing uses AI.

Finance uses AI.

Operations uses AI.

Human resources uses AI.

Sales uses AI.

Customer service uses AI.

Research and development uses AI.

Leadership itself is beginning to use AI as a thinking partner.

Unlike previous technologies that affected a particular department, AI has the potential to influence the entire enterprise.

This means leaders cannot delegate AI solely to the technology team.

The conversation belongs in every leadership meeting.

It belongs in strategic planning discussions.

It belongs in workforce planning.

It belongs in innovation conversations.

It belongs in discussions about organizational design.

AI is becoming a business issue, not merely a technology issue.

The Real Opportunity

Much of the public conversation about AI focuses on automation.

That focus is understandable, but it is incomplete.

The greatest opportunity may not be automation.

It may be augmentation.

Automation asks:

"How can machines perform tasks currently done by people?"

Augmentation asks:

"How can technology help people perform at higher levels?"

This distinction matters.

High-performance organizations do not simply seek to replace people.

They seek to increase the value people create.

Imagine a sales professional who can prepare for client meetings in minutes rather than hours.

Imagine a manager who receives immediate insights from thousands of data points.

Imagine a customer service representative who can resolve issues faster because AI instantly provides relevant information.

Imagine executives who can evaluate strategic alternatives more quickly and thoroughly.

In each case, AI enhances human performance rather than replacing it.

The most successful organizations will focus on amplification rather than elimination.

The Leadership Imperative

Every generation of leaders faces defining challenges.

For some, it was globalization.

For others, it was digital transformation.

For today's leaders, one of those defining challenges will be artificial intelligence.

The responsibility is not simply to adopt new technology.

The responsibility is to help people navigate change successfully.

Employees want clarity.

Customers want value.

Investors want results.

Communities want responsible leadership.

Leaders must balance all four.

That requires more than technical knowledge.

It requires vision.

It requires communication.

It requires courage.

It requires learning.

Most importantly, it requires the ability to help people see opportunity where others see uncertainty.

The organizations that thrive during periods of disruption are usually led by individuals who can create confidence without creating complacency.

They acknowledge the challenges.

They embrace the opportunities.

They move thoughtfully but decisively.

They learn continuously.

And they recognize that standing still is rarely a winning strategy.

Key Takeaways

Before moving to the next chapter, consider these five insights:

1. AI is not a future trend; it is a present reality.
2. The impact of AI extends far beyond technology departments.
3. AI should be viewed as an organizational transformation initiative, not merely a software implementation.
4. The greatest opportunity may be augmenting human capability rather than replacing human labor.
5. Leadership, culture, and organizational design will determine success far more than technology alone.

The age of AI has arrived.

The leaders who understand that fact, and act upon it thoughtfully, will help shape the next generation of high-performance organizations.

The next question is how.

That begins with understanding why AI is fundamentally changing the way organizations create value.

Next Chapter: Chapter 2 – Beyond Technology: Why AI Is an Organizational Transformation, Not an IT Project. This chapter builds the central argument for choosing the *"High-Performance Organizations"* transformational approach to thriving in the age of AI versus other *digitization only* approaches.

CHAPTER 2

Beyond Technology

Why AI Is an Organizational Transformation, Not an IT Project

One of the most expensive mistakes organizations make when implementing artificial intelligence is treating it as a technology project.

The pattern is familiar.

A leadership team hears about AI's potential. A task force is formed. Software is purchased. Consultants are hired. Pilot projects are launched. Employees receive a few hours of training.

Months later, the results are disappointing.

The technology works.

The organization does not.

Adoption is low.

Employees are confused.

Processes remain unchanged.

Managers continue making decisions the old way.

The anticipated productivity gains fail to materialize.

The investment produces far less value than expected.

The problem is rarely the technology itself.

The problem is that the organization attempted to add AI without changing how the organization thinks, operates, learns, and creates value.

In other words, they tried to install AI without transforming the organization.

The Difference Between Digitization and Transformation

Many leaders confuse digitization with transformation.

Digitization involves using technology to improve existing processes.

Transformation involves rethinking how work is performed and how value is created.

The difference is significant.

Imagine a company that replaces paper forms with electronic forms.

That is digitization.

Now imagine a company that completely redesigns the workflow, eliminates unnecessary approvals, automates routine decisions, provides real-time visibility, and empowers employees to focus on higher-value activities.

That is transformation.

AI creates opportunities for transformation because it changes what is possible.

Tasks that once required hours can be completed in minutes.

Analysis that once required teams can be performed instantly.

Information that once remained hidden can now be surfaced and acted upon.

But realizing these benefits requires leaders to rethink the organization itself.

Technology alone cannot accomplish that.

The Four Systems That Must Change

When organizations successfully implement AI, they rarely focus exclusively on technology.

Instead, they transform four interconnected systems:

System 1: The Work System

Organizations must reconsider how work gets done.

Many jobs contain activities that are repetitive, predictable, and data-intensive.

Other activities require judgment, creativity, emotional intelligence, negotiation, and leadership.

The goal is not to eliminate people.

The goal is to redesign work so that people spend more time doing what humans do best and less time doing what machines can do faster.

High-performance organizations constantly ask:

- What work should be automated?
- What work should be augmented?
- What work should remain entirely human?
- How should responsibilities evolve?

Organizations that answer these questions thoughtfully gain significant advantages in productivity and effectiveness.

System 2: The Decision-Making System

Many organizations struggle because decisions move too slowly.

Information sits in silos.

Approvals stack up.

Leaders spend weeks gathering information before acting.

AI offers an opportunity to accelerate decision-making.

However, faster information does not automatically create better decisions.

Organizations must redesign how decisions are made.

They must determine:

- Which decisions can be automated?
- Which decisions require human review?
- Which decisions require executive involvement?
- How should AI-generated recommendations be evaluated?

The organizations that thrive will combine machine intelligence with human judgment.

Neither alone is sufficient.

Together they can become extraordinarily powerful.

System 3: The Learning System

In the past, organizations could operate successfully for years with relatively stable knowledge and skills.

Those days are ending.

The AI era rewards learning organizations.

The most valuable employees will not necessarily be those with the most knowledge today.

They will be those who can learn fastest tomorrow.

Similarly, the most successful organizations will not necessarily be those with the largest budgets.

They will be those that learn, adapt, and innovate more rapidly than competitors.

Leaders must create environments where learning becomes a continuous process rather than an occasional event.

Training can no longer be viewed as something employees attend.

Learning must become part of how work is performed every day.

System 4: The Leadership System

Perhaps the greatest transformation occurs within leadership itself.

Traditional leadership models often relied on controlling information and directing activities.

AI changes that dynamic.

Information is becoming more accessible.

Insights are becoming more widely distributed.

Employees increasingly expect participation, transparency, and collaboration.

As a result, leaders must evolve.

Their role shifts from being the primary source of answers to becoming architects of performance.

They create clarity.

They align priorities.

They build trust.

They develop talent.

They create systems that allow people and technology to work together effectively.

This leadership evolution may be the most important transformation of all.

The Human Side of AI

Many discussions about AI focus on capabilities.

Far fewer discussions focus on emotions.

Yet emotions often determine whether transformation succeeds or fails.

Employees may wonder:

- Will AI replace my job?
- Will my skills remain relevant?
- Can I trust these systems?
- Will I be able to succeed in this new environment?

These concerns are natural.

Every major transformation creates uncertainty.

Leaders who ignore these concerns create resistance.

Leaders who address them create engagement.

The organizations that succeed recognize that AI implementation is not only a technology challenge.

It is a human challenge.

People need communication.

They need training.

They need support.

They need opportunities to contribute.

Most importantly, they need confidence that they have a place in the future being created.

Culture Eats Technology for Breakfast

Many executives have heard the saying:

"Culture eats strategy for breakfast."

The same principle applies to AI.

A weak culture can undermine even the most advanced technology investments.

Consider two organizations implementing identical AI tools.

The first organization has:

- High trust
- Strong communication
- A learning mindset
- Collaborative leadership
- Clear accountability

The second organization has:

- Fear of change
- Poor communication
- Low trust
- Political behavior
- Resistance to innovation

Which organization is likely to achieve better results?

The answer is obvious.

Technology amplifies existing organizational strengths and weaknesses.

Organizations with healthy cultures often accelerate.

Organizations with dysfunctional cultures often expose those dysfunctions more rapidly.

Before leaders ask whether they are ready for AI, they should ask whether their culture is ready for change.

The New Organizational Design Challenge

Historically, organizational design focused on optimizing people, processes, and resources.

Today, leaders must design organizations that integrate:

- Human intelligence
- Artificial intelligence
- Continuous learning
- Rapid adaptation
- Innovation at scale

This requires a new mindset.

The goal is no longer efficiency alone.

The goal is adaptability.

Organizations must become capable of evolving continuously.

The winners in the AI era will not necessarily be the largest organizations.

They will be the organizations that learn fastest, adapt fastest, and create value fastest.

A Leadership Reflection

Consider your own organization.

If AI disappeared tomorrow, would your organization still struggle with:

- Slow decision-making?
- Poor communication?
- Siloed departments?
- Resistance to change?
- Lack of accountability?

- Weak leadership alignment?

If the answer is yes, AI will not solve those problems.

In fact, it may expose them.

The organizations that gain the greatest value from AI first strengthen the organizational foundations that support high performance.

They understand a simple truth:

Technology can accelerate performance.

But it cannot substitute for leadership.

Key Takeaways

1. AI implementation is an organizational transformation initiative, not merely a technology project.
2. Successful adoption requires changes in work systems, decision-making systems, learning systems, and leadership systems.
3. Employee trust and engagement are critical to successful transformation.
4. Culture often determines whether AI investments succeed or fail.
5. High-performance organizations redesign themselves to integrate human and artificial intelligence effectively.

The leaders who view AI as a software purchase will likely be disappointed.

The leaders who view AI as an opportunity to redesign their organizations for the future will unlock its greatest potential.

The next step is understanding how AI is changing the competitive landscape itself, and why the rules of business are being rewritten in real time.

Next Chapter: Chapter 3 , The New Rules of Competition: How AI Is Reshaping Markets, Industries, and Business Models. This chapter will help you and your leadership team understand why AI is becoming a strategic necessity rather than a discretionary investment.

CHAPTER 3

The New Rules of Competition

How AI Is Reshaping Markets, Industries, and Business Models

For decades, organizations competed using a familiar set of advantages.

They sought better products.

Better locations.

Better talent.

More capital.

Stronger brands.

More efficient operations.

These factors still matter.

But artificial intelligence is changing the competitive landscape in ways many leaders have yet to fully appreciate.

The rules of competition are being rewritten.

The organizations that understand these new rules will create extraordinary advantages.

Those that fail to adapt may find themselves competing against organizations that are faster, smarter, more agile, and more capable of creating value at scale.

The challenge is that many leaders continue operating with assumptions that were formed in a pre-AI world.

The marketplace has moved on.

Speed Has Become a Competitive Weapon

Historically, organizations often had months or even years to react to changing market conditions.

Today, information moves instantly.

Customer expectations evolve rapidly.

Competitors launch innovations faster than ever.

AI accelerates this reality.

A marketing campaign that once required weeks can now be developed in hours.

A competitive analysis that once took days can be completed in minutes.

A strategic report that required multiple meetings can often be generated almost immediately.

Speed alone does not guarantee success.

However, the ability to learn, decide, and act faster than competitors is becoming one of the most powerful advantages an organization can possess.

In many industries, the winner is no longer the largest organization.

The winner is often the organization that learns fastest.

Information Is No Longer Scarce

For most of business history, information was a source of power because it was limited.

Executives had access to information others did not.

Organizations invested heavily in collecting data.

The challenge was obtaining information.

Today, the challenge is making sense of it.

Organizations are drowning in data while starving for insight.

Artificial intelligence changes this equation.

AI can analyze vast amounts of information, identify patterns, detect anomalies, and surface insights at speeds impossible for human teams alone.

This creates a significant shift.

Competitive advantage increasingly comes not from possessing information but from interpreting and acting on information more effectively than competitors.

The organizations that win will not necessarily have the most data.

They will have the greatest ability to convert data into decisions and decisions into results.

Customer Expectations Are Changing

Customers may not always know how AI works.

They do know when experiences improve.

Every time customers experience faster service, personalized recommendations, immediate answers, or seamless interactions, their expectations rise.

What once felt exceptional quickly becomes expected.

Consider how customer expectations have evolved regarding:

- Online shopping
- Food delivery
- Banking
- Travel booking
- Entertainment streaming
- Customer service

Convenience has become the standard.

Speed has become the expectation.

Personalization has become the norm.

Organizations that fail to meet these expectations risk becoming irrelevant.

The AI era is creating customers who expect businesses to know them, understand them, and respond quickly to their needs.

Organizations that deliver these experiences consistently gain significant competitive advantages.

The Democratization of Capability

One of the most profound changes AI creates is the democratization of capability.

Historically, large organizations enjoyed advantages because they possessed resources that smaller organizations could not access.

They could hire more specialists.

Purchase more technology.

Conduct more research.

Build larger support teams.

Artificial intelligence is beginning to level portions of the playing field.

A small business can now access capabilities that once required entire departments.

An entrepreneur can leverage tools that previously existed only inside large enterprises.

A startup can compete more effectively against established competitors.

This does not mean size no longer matters.

It does mean that intelligence, adaptability, and execution are becoming increasingly important.

The gap between what small organizations can accomplish and what large organizations can accomplish is narrowing.

This creates tremendous opportunities, and new threats.

The Rise of the Augmented Workforce

The workforce itself is evolving.

Historically, organizations added capacity primarily by adding people.

Need more output?

Hire more employees.

Need more analysis?

Build a larger team.

Need more support?

Expand the department.

AI introduces a different possibility.

Organizations can increase capacity by augmenting existing employees.

One highly skilled professional supported by AI may accomplish what previously required multiple people.

This does not eliminate the need for talent.

In many cases, it increases the value of talented individuals.

Employees who know how to leverage AI effectively become dramatically more productive.

As a result, organizations may increasingly compete based on the quality of their talent and the quality of the tools supporting that talent.

The future belongs not simply to organizations with great people.

It belongs to organizations with great people empowered by great systems.

Innovation Is Accelerating

Innovation has always been important.

Now it is becoming essential.

AI reduces many of the barriers that historically slowed innovation.

Ideas can be tested faster.

Research can be conducted more efficiently.

Prototypes can be developed more quickly.

Customer feedback can be analyzed almost instantly.

As a result, innovation cycles are shortening.

Organizations that previously launched new products annually may now innovate quarterly.

Organizations that previously innovated quarterly may now innovate continuously.

This acceleration creates pressure.

Leaders can no longer rely solely on what worked yesterday.

They must create organizations capable of learning and adapting constantly.

The ability to innovate repeatedly is becoming more valuable than any single innovation.

Business Models Are Being Reinvented

Every major technological shift creates new business models.

Artificial intelligence is no exception.

Organizations are discovering opportunities to:

- Deliver services differently
- Create new revenue streams
- Personalize offerings
- Automate value delivery
- Expand market reach
- Reduce operating costs
- Improve customer outcomes

Some organizations will use AI to improve existing business models.

Others will create entirely new ones.

History suggests that disruptive competitors often emerge from unexpected places.

The greatest threat to an industry leader is not always another industry leader.

Sometimes it is a small, agile organization willing to rethink the rules entirely.

Leaders must ask themselves:

If we were starting our organization today, knowing what AI makes possible, would we design our business the same way?

The answer to that question often reveals opportunities hiding in plain sight.

The Strategic Advantage of Learning

As AI continues evolving, one reality becomes increasingly clear:

No organization has all the answers.

Technology changes too quickly.

Markets evolve too rapidly.

Customer expectations shift too often.

The organizations that succeed will not be those that know everything.

They will be those that learn everything they need to know faster than competitors.

Learning becomes strategy.

Experimentation becomes strategy.

Adaptability becomes strategy.

Curiosity becomes strategy.

Organizations that create cultures of continuous learning will consistently outperform organizations that cling to yesterday's assumptions.

The Emerging Winners

As we look across industries, a pattern is beginning to emerge.

The organizations creating the greatest value in the AI era tend to share several characteristics:

They learn rapidly.

They embrace experimentation.

They empower employees.

They make decisions quickly.

They use data intelligently.

They communicate clearly.

They maintain strong cultures.

They focus relentlessly on customer value.

Most importantly, they view AI as a strategic capability rather than a technological novelty.

They understand that AI is not the destination.

It is an accelerator.

The destination remains the same as it has always been:

Creating value for customers, employees, investors, and society.

Leadership Questions

As you consider your organization, reflect on these questions:

- Where is AI likely to create the greatest competitive advantage in our industry?
- How quickly are customer expectations changing?
- What capabilities could AI help us develop?
- What assumptions about our business may no longer be true?
- How can we increase the speed of learning across our organization?
- If a startup entered our market tomorrow using AI aggressively, how would it compete against us?

The answers may reveal opportunities and vulnerabilities that deserve immediate attention.

Key Takeaways

1. AI is changing the rules of competition across nearly every industry.
2. Speed of learning and adaptation is becoming a major competitive advantage.
3. Customer expectations are rising as AI-enhanced experiences become common.
4. AI is democratizing capabilities that were once available only to large organizations.
5. Organizations must continuously rethink how they create value and compete.

6. The future belongs to organizations that combine human talent, intelligent systems, and a culture of continuous learning.

The organizations that thrive in the coming decade will not necessarily be the ones with the most advanced technology.

They will be the ones that learn fastest, adapt fastest, and create value fastest.

And that raises an important question:

If AI can analyze information, automate tasks, and accelerate decisions, what role remains for human beings?

The answer may be the most important lesson in this entire book.

Next Chapter: Chapter 4 , The Human Opportunity: What Humans Will Do Better Than Ever in an AI World. Understanding the *Key Takeaways* within this chapter can become a powerful bridge from technology into leadership, culture, creativity, judgment, wisdom, relationships and the uniquely human capabilities that will increase in value rather than decrease in the age of AI.

IN THE
AGE
OF
AI

CHAPTER 4

The Human Opportunity

What Humans Will Do Better Than Ever in an AI World

Whenever a transformational technology emerges, people naturally ask the same question:

"What happens to us?"

The question is understandable.

Will jobs disappear?

Will machines replace people?

Will human skills become less valuable?

Will artificial intelligence eventually outperform humans in every area?

These concerns dominate many conversations about AI.

Yet history suggests we may be asking the wrong question.

The more important question is this:

What becomes possible for human beings when intelligent machines handle more of the routine work?

This question shifts the focus from fear to opportunity.

It moves the conversation from replacement to enhancement.

And it reveals one of the most important truths of the AI era:

As artificial intelligence becomes more capable, uniquely human capabilities become more valuable.

The Myth of Human Replacement

Throughout history, new technologies have altered jobs, industries, and economic systems.

The automobile changed transportation.

The computer changed office work.

The internet changed communication and commerce.

Each innovation eliminated certain tasks while creating entirely new opportunities.

Artificial intelligence is likely to follow a similar pattern.

Some tasks will be automated.

Some jobs will evolve.

Some roles will disappear.

New roles will emerge.

But most organizations are not struggling because they have too many talented people.

They are struggling because talented people spend too much time doing work that does not fully utilize their abilities.

Highly educated professionals spend hours searching for information.

Managers spend excessive time creating reports.

Salespeople spend significant time preparing administrative materials.

Executives spend countless hours gathering data before making decisions.

AI has the potential to reduce these burdens.

When that happens, people can focus more energy on the activities that create the greatest value.

The Rise of Human Intelligence

Ironically, the age of artificial intelligence may become one of the greatest opportunities for human intelligence.

Not intelligence measured solely by IQ.

But intelligence expressed through judgment, wisdom, creativity, empathy, leadership, and innovation.

Consider the difference.

Artificial intelligence can provide answers.

Human intelligence determines which questions matter.

Artificial intelligence can identify patterns.

Human intelligence determines which patterns are meaningful.

Artificial intelligence can generate options.

Human intelligence determines which option aligns with values, ethics, and long-term goals.

Artificial intelligence can process information.

Human intelligence provides wisdom.

The distinction is critical.

Organizations that understand this will create tremendous advantages.

Organizations that ignore it may unintentionally weaken the very capabilities that make them successful.

Creativity Becomes More Valuable

Many people assume creativity belongs exclusively to artists, writers, musicians, and designers.

In reality, creativity is essential to every profession.

Entrepreneurs use creativity to identify opportunities.

Leaders use creativity to solve problems.

Managers use creativity to improve systems.

Sales professionals use creativity to connect with customers.

Engineers use creativity to design solutions.

Innovation itself is an act of creativity.

Artificial intelligence can generate ideas.

It can create drafts.

It can suggest alternatives.

But creativity is more than generating possibilities.

Creativity involves imagination.

It involves vision.

It involves seeing opportunities others overlook.

The organizations that thrive will encourage employees to use AI as a creative partner while continuing to develop their own creative thinking abilities.

The goal is not to outsource creativity.

The goal is to amplify it.

Judgment Becomes a Strategic Asset

One of the greatest misconceptions about AI is that information automatically leads to good decisions.

In reality, organizations rarely suffer from a lack of information.

They suffer from poor judgment.

Judgment involves evaluating trade-offs.

Balancing competing priorities.

Considering long-term consequences.

Recognizing context.

Understanding people.

Navigating uncertainty.

These capabilities remain deeply human.

Two leaders can receive identical information and make entirely different decisions.

The difference is judgment.

As AI provides more information and more recommendations, the quality of human judgment becomes increasingly important.

The future belongs not to leaders who know the most.

It belongs to leaders who exercise the best judgment.

Empathy Creates Differentiation

Machines can analyze sentiment.

They can recognize emotional patterns.

They can simulate conversations.

But empathy remains fundamentally human.

Empathy is the ability to understand another person's experience.

It allows leaders to build trust.

It helps managers resolve conflict.

It enables sales professionals to understand customer needs.

It strengthens teamwork.

It improves collaboration.

As routine interactions become increasingly automated, genuine human connection becomes more valuable.

Customers remember how they were treated.

Employees remember how leaders made them feel.

Teams perform better when people feel understood and respected.

Organizations that preserve and strengthen empathy will create experiences competitors struggle to replicate.

Leadership Matters More Than Ever

Some people assume AI will reduce the importance of leadership.

The opposite is likely true.

Periods of uncertainty increase the need for leadership.

Employees want direction.

Customers want confidence.

Stakeholders want clarity.

Communities want responsible decision-making.

Technology can provide information.

Leadership provides meaning.

Technology can offer recommendations.

Leadership creates alignment.

Technology can automate processes.

Leadership inspires commitment.

As organizations navigate AI-driven transformation, leaders must help people understand not only what is changing but why it matters.

This requires communication.

Vision.

Trust.

Credibility.

And courage.

The future will not require less leadership.

It will require better leadership.

Wisdom: The Ultimate Human Advantage

Knowledge and wisdom are not the same thing.

Knowledge tells us what can be done.

Wisdom helps us determine what should be done.

Artificial intelligence is expanding access to knowledge at an extraordinary pace.

That makes wisdom even more important.

Organizations will increasingly face questions such as:

- Should we automate this process?
- Should we collect this data?
- Should we use this technology?
- Should we make this decision?

These questions cannot be answered by algorithms alone.

They require values.

Ethics.

Perspective.

Experience.

Wisdom.

The leaders who develop wisdom will become increasingly valuable in an age overflowing with information.

The Human Skills That Will Define the Future

As AI becomes more integrated into organizational life, several human capabilities will grow in importance.

Critical Thinking

The ability to evaluate information, challenge assumptions, and make sound decisions.

Communication

The ability to create clarity, influence others, and build understanding.

Collaboration

The ability to work effectively with diverse teams and perspectives.

Emotional Intelligence

The ability to understand and manage emotions while building strong relationships.

Adaptability

The ability to learn, evolve, and thrive amid change.

Creativity

The ability to imagine new possibilities and solve problems innovatively.

Leadership

The ability to inspire, align, and mobilize people toward meaningful goals.

These skills are not being replaced by AI.

They are becoming more valuable because of AI.

The New Partnership

The future is not a competition between humans and machines.

It is a partnership.

The most successful organizations will not ask:

"What can humans do?"

Or:

"What can AI do?"

Instead, they will ask:

"What can humans and AI accomplish together that neither could accomplish alone?"

That question creates a fundamentally different mindset.

It encourages collaboration rather than fear.

Possibility rather than limitation.

Innovation rather than resistance.

High-performance organizations will build systems where human strengths and artificial intelligence complement one another.

The result will be higher levels of performance, creativity, learning, and value creation.

A Leadership Reflection

Think about your highest-performing employees.

Why are they successful?

Is it because they can process information faster than everyone else?

Probably not.

More likely, they excel because they:

- Build relationships.
- Earn trust.
- Solve problems.
- Exercise judgment.
- Communicate effectively.
- Adapt to change.
- Inspire others.
- Create value.

These capabilities will remain valuable regardless of how advanced technology becomes.

In fact, they may become even more valuable.

Key Takeaways

1. AI is changing tasks more rapidly than it is changing the need for human capability.
2. Creativity, judgment, empathy, and wisdom are becoming increasingly valuable.
3. Human intelligence and artificial intelligence are most powerful when combined.
4. Leadership becomes more important during periods of technological transformation.
5. Organizations should focus on augmenting human capability rather than replacing it.
6. The future belongs to organizations that develop both technological excellence and human excellence.

The organizations that thrive in the AI era will not simply possess powerful technologies.

They will possess powerful cultures.

They will develop exceptional people.

They will create environments where learning, innovation, and leadership flourish.

And that raises the next challenge:

How do leaders intentionally build organizations capable of thriving in this new environment?

The answer begins with purpose.

Before organizations implement AI, redesign processes, or invest in technology, they must first understand what they are trying to accomplish and why.

Without purpose, technology becomes noise.

With purpose, technology becomes a force multiplier.

Next Chapter: Chapter 5 , Start with Purpose: Aligning AI Initiatives with Strategy and Mission. This chapter begins Part II and introduces a practical framework leaders can use to decide where AI should, and should not, be deployed.

IN THE
AGE
OF
AI

PART II

BUILDING THE HIGH-PERFORMANCE AI ORGANIZATION

IN THE
AGE
OF
AI

CHAPTER 5

Start with Purpose

Aligning AI Initiatives with Strategy and Mission

One of the fastest ways to waste money on artificial intelligence is to begin with the technology.

Unfortunately, that is exactly where many organizations start.

A competitor announces an AI initiative.

A board member asks about AI strategy.

An article highlights impressive productivity gains.

A vendor demonstrates a powerful new platform.

Suddenly, leaders feel pressure to act.

They begin evaluating tools, exploring vendors, and launching projects.

The excitement is understandable.

The problem is that they are asking the wrong question.

The first question should never be:

"What AI tools should we buy?"

The first question should always be:

"What are we trying to accomplish?"

High-performance organizations begin with purpose.

They begin with strategy.

They begin with mission.

Only then do they determine how technology can help them achieve those objectives.

Technology should support the mission.

The mission should never be redesigned around the technology.

The AI Gold Rush

Throughout history, every major innovation has created a period of excitement that resembles a gold rush.

Organizations rush to participate.

Investors rush to invest.

Vendors rush to sell.

Consultants rush to advise.

Artificial intelligence is no exception.

Organizations are launching AI initiatives across every industry.

Some are generating significant value.

Others are spending substantial resources with little to show for it.

The difference is rarely the technology.

The difference is clarity.

Successful organizations know why they are implementing AI.

Unsuccessful organizations implement AI because everyone else appears to be doing it.

One approach is strategic.

The other is reactive.

High-performance organizations avoid chasing technology trends.

Instead, they focus relentlessly on business outcomes.

The Purpose Filter

Before approving any AI initiative, leaders should apply what I call the Purpose Filter.

Ask five simple questions:

Question 1:

What organizational objective are we trying to achieve?

Examples might include:

- Improving customer experience
- Increasing productivity
- Reducing costs
- Accelerating innovation
- Improving decision quality
- Enhancing employee engagement
- Expanding revenue

If the objective is unclear, the initiative is likely premature.

Question 2:

How will success be measured?

Many organizations invest in technology without defining success.

As a result, they struggle to determine whether the investment created value.

Metrics create clarity.

Without metrics, opinions dominate.

Question 3:

Who benefits?

Customers?

Employees?

Managers?

Investors?

Partners?

Communities?

The most successful AI initiatives create meaningful value for multiple stakeholders.

Question 4:

What problem are we solving?

Technology should solve problems.

It should not create projects for the sake of having projects.

Clear problems create clear priorities.

Question 5:

Does this align with our mission and values?

Not everything that can be automated should be automated.

Not everything that can be measured should be measured.

Not everything that can be implemented should be implemented.

Mission and values provide guidance.

Organizations that lose sight of their purpose often create unintended consequences.

Purpose Creates Focus

One of the greatest leadership responsibilities is deciding what not to pursue.

Resources are limited.

Time is limited.

Attention is limited.

Every organization faces more opportunities than it can realistically pursue.

Artificial intelligence multiplies those opportunities.

Without focus, organizations become distracted.

They launch too many initiatives.

They spread resources too thin.

They confuse employees.

They create implementation fatigue.

Purpose helps leaders focus.

It answers the question:

"What matters most?"

Organizations with clear purpose make better decisions because they understand their priorities.

They can evaluate opportunities through a consistent lens.

The Mission Multiplier

When properly aligned, AI becomes a mission multiplier.

Consider a healthcare organization whose mission is improving patient outcomes.

AI can help:

- Identify health risks earlier.
- Improve treatment recommendations.
- Streamline administrative work.
- Increase access to care.

The technology supports the mission.

Now consider an educational institution whose mission is helping students learn.

AI can:

- Personalize learning experiences.
- Identify struggling students sooner.
- Improve curriculum development.
- Reduce administrative burdens on teachers.

Again, the technology supports the mission.

The same principle applies to businesses, nonprofits, government agencies, and community organizations.

The purpose comes first.

Technology amplifies the purpose.

Why Employees Resist

Many AI initiatives fail because leaders communicate technology rather than purpose.

Employees hear:

"We're implementing AI."

That statement often triggers questions.

Why?

What does this mean for me?

Will my job change?

Will I still be valuable?

Can I succeed in this new environment?

Now consider a different message:

"We're implementing AI so we can reduce administrative work, serve customers more effectively, and allow employees to spend more time doing meaningful, high-value work."

Notice the difference.

The technology remains the same.

The purpose changes everything.

People support what they understand.

People commit to what they believe in.

Purpose creates understanding.

Purpose creates belief.

Purpose creates engagement.

The Three Levels of Alignment

High-performance organizations align AI initiatives at three levels.

Strategic Alignment

Does the initiative support long-term organizational objectives?

If the initiative does not contribute to strategic priorities, its value should be questioned.

Operational Alignment

Does the initiative improve how work is performed?

Can it improve efficiency, effectiveness, quality, or customer outcomes?

Operational improvements create tangible value.

Human Alignment

Does the initiative help people succeed?

Does it make work better?

Does it improve experiences?

Does it support growth and development?

Organizations that achieve all three forms of alignment generate significantly greater returns on their AI investments.

The AI Strategy Framework

When evaluating opportunities, leaders should consider five categories where AI can create value.

Productivity

Helping people complete work more efficiently.

Decision-Making

Helping leaders make better-informed decisions.

Customer Experience

Improving service, responsiveness, and personalization.

Innovation

Accelerating creativity, experimentation, and problem-solving.

Growth

Creating new products, services, markets, and revenue streams.

Nearly every AI initiative can be linked to one or more of these categories.

If it cannot, leaders should question its strategic importance.

A Leadership Lesson

Throughout my career, I have observed organizations become enamored with tools.

New software.

New systems.

New methodologies.

New technologies.

The pattern is always the same.

Organizations become excited about the possibility of improvement.

Yet the organizations that consistently outperform their peers rarely obsess over tools.

They obsess over outcomes.

They ask:

What are we trying to accomplish?

How will we know if we succeed?

How can we create more value?

Technology is simply one of many tools available to help answer those questions.

AI is powerful.

But purpose remains more powerful.

Purpose determines direction.

Technology determines speed.

An organization moving quickly in the wrong direction still ends up in the wrong place.

Leadership Reflection

Consider your organization.

Can every member of your leadership team clearly answer:

- Why are we exploring AI?
- What outcomes are we pursuing?
- How will success be measured?
- How does this support our mission?
- How does this create value for customers and employees?

If those questions cannot be answered clearly, the conversation should begin there.

Clarity precedes execution.

Purpose precedes transformation.

Key Takeaways

1. Organizations should begin AI initiatives with purpose, not technology.
2. AI is most valuable when aligned with mission, strategy, and organizational objectives.
3. The Purpose Filter helps leaders evaluate opportunities effectively.
4. Employees are more likely to support change when they understand its purpose.
5. Strategic, operational, and human alignment are all necessary for success.
6. Technology should amplify mission rather than define it.

The organizations that achieve extraordinary results with AI do not start by asking what technology can do.

They start by asking what their organization is called to achieve.

Once purpose is clear, the next challenge becomes designing an organization capable of delivering that purpose consistently in an AI-enabled world.

That requires a fresh look at organizational structure, decision-making, and the architecture of performance itself.

Next Chapter: Chapter 6 , Designing the AI-Ready Organization: Structure, Roles, and Decision-Making for the Future. This chapter introduces a practical organizational design model that executives can use to redesign their companies for the AI era.

IN THE
AGE
OF
AI

CHAPTER 6

Designing the AI-Ready Organization

Structure, Roles, and Decision-Making for the Future

Most organizations today were designed for a world that no longer exists.

They were built for slower communication.

Slower decision-making.

Slower innovation cycles.

More predictable markets.

Information flowed upward through layers of management. Decisions flowed downward through chains of command. Departments often operated independently, and expertise was concentrated in a relatively small number of individuals.

For decades, this approach worked reasonably well.

The AI era changes the equation.

Information now moves instantly.

Customers expect rapid responses.

Competitors innovate continuously.

Employees have access to unprecedented levels of knowledge.

Artificial intelligence is accelerating all of these trends.

As a result, leaders face a critical question:

Is our organization designed for the future we are entering, or the past we are leaving?

The answer may determine whether the organization thrives or struggles in the years ahead.

Organizational Design Is Not an Org Chart

When many leaders hear the term "organizational design," they immediately think about organizational charts.

Boxes.

Titles.

Reporting relationships.

While those elements matter, they represent only a small portion of organizational design.

High-performance organizations are built upon five interconnected components:

1. Purpose
2. Structure
3. Processes
4. People
5. Culture

The organizational chart is merely a visual representation of how some of these components interact.

True organizational design focuses on how work flows, how decisions are made, how information is shared, and how value is created.

The AI-ready organization must be designed with all five components in mind.

From Hierarchies to Networks

Traditional organizations often resemble pyramids.

Information flows upward.

Instructions flow downward.

Each level waits for approval from the level above.

This structure can create stability.

It can also create delays.

In a rapidly changing environment, delays become costly.

AI-ready organizations increasingly operate like networks.

Information moves more freely.

Employees have access to better data.

Cross-functional collaboration becomes easier.

Decision-making occurs closer to where the work is being performed.

Leaders still provide direction and accountability, but they avoid becoming bottlenecks.

The goal is not eliminating hierarchy.

The goal is eliminating unnecessary friction.

Organizations that reduce friction gain speed.

Organizations that gain speed gain competitive advantage.

The New Role of Leadership

As information becomes more accessible, the role of leadership evolves.

Historically, managers often acted as information gatekeepers.

They collected information, interpreted it, and communicated it to others.

Today, much of that information is available instantly.

Employees can access reports, dashboards, market intelligence, and AI-generated insights in seconds.

This changes the leader's role.

Leaders become architects rather than gatekeepers.

They create alignment.

They establish priorities.

They define standards.

They build culture.

They develop people.

They ensure that organizational energy is directed toward the most important objectives.

The future leader spends less time controlling information and more time creating clarity.

Designing Around Decisions

One of the most powerful ways to improve organizational performance is to redesign decision-making.

Many organizations unknowingly create decision bottlenecks.

Employees wait for approvals.

Managers wait for meetings.

Departments wait for consensus.

Projects stall while decisions move slowly through the organization.

Artificial intelligence can accelerate information gathering, but leaders must still redesign decision processes.

Consider these questions:

- Which decisions can be automated?
- Which decisions should be delegated?
- Which decisions require human judgment?
- Which decisions require executive involvement?

High-performance organizations intentionally answer these questions.

They create decision frameworks that balance speed, quality, and accountability.

The result is an organization that can respond more quickly to changing conditions.

The Rise of Human-AI Teams

The future organization will increasingly rely on teams composed of both humans and AI systems.

This concept requires a shift in thinking.

Many leaders still view AI as a tool.

While that is technically accurate, it may not be the most useful perspective.

A better perspective is to view AI as a capability multiplier.

Just as a team benefits from specialists with different strengths, organizations can benefit from AI systems that contribute specific capabilities.

Humans bring:

- Creativity
- Judgment
- Empathy
- Leadership
- Ethics
- Relationship-building

AI contributes:

- Speed
- Analysis
- Pattern recognition
- Automation
- Information retrieval
- Scalability

The goal is not determining which is better.

The goal is designing systems that maximize the strengths of both.

Organizations that learn to do this effectively will outperform those that rely exclusively on either humans or technology.

Building for Adaptability

For much of the twentieth century, efficiency was considered the primary objective of organizational design.

Efficiency remains important.

However, adaptability may now be even more important.

An efficient organization that cannot adapt becomes vulnerable.

A slightly less efficient organization that learns quickly often gains long-term advantage.

This means AI-ready organizations must be designed for flexibility.

They must be capable of:

- Learning rapidly
- Experimenting safely
- Adapting continuously
- Sharing knowledge effectively
- Responding to changing conditions

Leaders should ask themselves:

How quickly can our organization learn?

How quickly can we change?

How quickly can we act?

The answers to these questions increasingly determine organizational success.

The AI-Ready Workforce

Organizational design is not only about systems and structures.

It is also about people.

Many organizations make the mistake of assuming AI readiness is primarily a technology issue.

In reality, workforce readiness may be even more important.

Employees need:

- New skills
- New mindsets
- New tools
- New ways of working

More importantly, they need confidence.

They need to understand where they fit in the future organization.

The most successful organizations invest heavily in helping employees adapt and grow.

They view workforce development as a strategic investment rather than an administrative expense.

They recognize that technology evolves quickly.

People must evolve alongside it.

The Five Characteristics of AI-Ready Organizations

As organizations prepare for the future, five characteristics consistently appear among high performers.

1. Clarity

Everyone understands the mission, priorities, and desired outcomes.

2. Agility

The organization can adapt quickly to changing circumstances.

3. Collaboration

Departments work together rather than operating in silos.

4. Learning

Continuous improvement becomes part of daily operations.

5. Accountability

Individuals and teams take ownership of results.

These characteristics are not created by technology.

They are created by leadership.

Technology simply amplifies them.

Designing for Scale

One of the greatest benefits of AI is scalability.

Organizations can often serve more customers, process more information, and generate more value without proportionally increasing resources.

However, scalability requires intentional design.

Poorly designed processes scale inefficiency.

Poor communication scales confusion.

Weak accountability scales dysfunction.

Technology magnifies whatever already exists.

This is why organizational design matters so much.

AI can help organizations scale excellence.

It can also help organizations scale problems.

Leaders must decide which one they want to amplify.

A Leadership Reflection

Imagine you were starting your organization from scratch today.

You know what AI can do.

You understand how markets are changing.

You recognize how employees want to work.

You see how customer expectations are evolving.

Would you design the organization exactly as it exists today?

Or would you create something different?

Most leaders answer that question honestly and realize significant changes are needed.

That realization is not a criticism of the past.

It is an opportunity to prepare for the future.

The strongest organizations are not those that preserve every historical practice.

They are those that continually redesign themselves to remain effective.

Key Takeaways

1. Most organizations were designed for a pre-AI world and require thoughtful redesign.
2. Organizational design involves purpose, structure, processes, people, and culture, not just reporting relationships.

3. AI-ready organizations reduce friction and accelerate decision-making.
4. Leaders are evolving from information gatekeepers to architects of performance.
5. Human-AI teams will become a defining feature of high-performance organizations.
6. Adaptability is becoming as important as efficiency.
7. Workforce readiness is essential to organizational readiness.
8. Technology amplifies existing strengths and weaknesses, making organizational design more important than ever.

The future belongs to organizations that can learn faster, adapt faster, and execute faster than their competitors.

But no amount of organizational redesign will succeed without the right culture.

Culture ultimately determines whether people embrace change, resist change, or lead change.

And in the age of AI, culture may become the single most important competitive advantage an organization can possess.

Next Chapter: Chapter 7 , Creating a Culture of Innovation and Learning: How High-Performance Organizations Adapt Faster.

This chapter introduces practical frameworks for building a culture that embraces experimentation, continuous learning, accountability, psychological safety, and AI adoption, turning culture into a strategic advantage.

IN THE
AGE
OF
AI

CHAPTER 7

Creating a Culture of Innovation and Learning

How High-Performance Organizations Adapt Faster

If strategy determines where an organization is going, culture determines how fast it gets there.

Many leaders underestimate the power of culture because they cannot see it.

They can see buildings.

They can see technology.

They can see financial reports.

They can see organizational charts.

Culture is different.

Culture is invisible until it shows up in behavior.

It appears in how people communicate.

How they make decisions.

How they respond to mistakes.

How they handle conflict.

How they embrace, or resist, change.

And in the age of AI, culture may be the greatest competitive advantage an organization can possess.

Why?

Because technology is increasingly available to everyone.

Your competitors can purchase similar software.

They can access similar tools.

They can hire similar consultants.

What they cannot easily replicate is your culture.

Culture is the operating system of the organization.

When the operating system is healthy, everything performs better.

When the operating system is unhealthy, even the best technology struggles to create value.

The New Competitive Advantage

For years, organizations competed primarily on products, services, and operational efficiency.

Today, they increasingly compete on adaptability.

The ability to learn faster.

The ability to innovate faster.

The ability to respond faster.

The ability to improve faster.

Adaptability is no longer optional.

It has become essential.

Artificial intelligence accelerates change.

New tools appear constantly.

Customer expectations evolve rapidly.

Market conditions shift unexpectedly.

Organizations that cannot adapt quickly find themselves falling behind.

The question becomes:

How do organizations create cultures that embrace learning and innovation rather than resist them?

The answer begins with leadership.

Innovation Is Not a Department

One of the most common mistakes organizations make is treating innovation as a specialized function.

They create an innovation team.

A research department.

A transformation office.

A digital lab.

While these groups can be valuable, innovation cannot be delegated.

High-performance organizations recognize that innovation is everyone's responsibility.

Every employee sees problems.

Every employee encounters inefficiencies.

Every employee has ideas.

The challenge is creating an environment where those ideas are welcomed, explored, and acted upon.

Innovation should not be limited to developing new products.

Innovation includes:

- Improving processes
- Solving customer problems
- Increasing efficiency
- Enhancing collaboration
- Improving quality
- Reducing waste
- Creating new opportunities

Organizations that unleash innovation throughout the workforce gain a tremendous advantage.

Learning as a Strategic Capability

For much of history, learning occurred primarily through formal education.

Employees attended classes.

Managers participated in workshops.

Organizations conducted occasional training programs.

Those approaches remain useful.

But they are no longer sufficient.

In the AI era, learning must become continuous.

Knowledge changes too quickly.

Skills evolve too rapidly.

New opportunities emerge too frequently.

The organizations that thrive create systems that encourage learning every day.

Employees learn from:

- Colleagues
- Customers
- Experiments
- Data
- Feedback
- Reflection
- AI-powered tools

Learning becomes part of the work itself.

Not an activity separate from work.

Organizations that adopt this mindset create a powerful competitive advantage.

The Fear Factor

Despite all the benefits of innovation and learning, many employees resist change.

Not because they dislike improvement.

Because they fear uncertainty.

Questions begin to emerge:

What if I fail?

What if I make a mistake?

What if I don't understand the new technology?

What if I become irrelevant?

What if my idea sounds foolish?

These concerns are normal.

They are human.

The challenge for leaders is creating an environment where people feel safe enough to learn.

This is where psychological safety becomes critical.

Psychological Safety and Performance

Psychological safety is the belief that individuals can speak up, ask questions, offer ideas, and admit mistakes without fear of embarrassment or punishment.

Organizations with high psychological safety tend to:

- Learn faster
- Innovate more frequently
- Solve problems more effectively
- Collaborate more successfully
- Adapt more rapidly

This does not mean lowering standards.

In fact, high-performance cultures combine two seemingly opposite ideas:

High standards.

High support.

People are expected to perform.

They are also encouraged to learn.

Mistakes are examined rather than hidden.

Feedback is viewed as a gift rather than a threat.

Questions are encouraged rather than discouraged.

This environment accelerates growth.

From Knowing to Learning

Many organizations unknowingly reward certainty.

Managers are expected to have answers.

Leaders are expected to know.

Employees are expected to avoid mistakes.

While well-intentioned, this mindset can hinder innovation.

The future belongs to learning organizations.

Learning organizations recognize that no one has all the answers.

Curiosity becomes more valuable than certainty.

Questions become more valuable than assumptions.

Experimentation becomes more valuable than perfection.

In a rapidly changing environment, the willingness to learn often matters more than existing knowledge.

The organizations that embrace this reality outperform those that cling to the illusion of certainty.

The Experimentation Mindset

Innovation requires experimentation.

Not every idea will succeed.

Not every initiative will generate results.

Not every pilot program will achieve its objectives.

This is normal.

The goal is not avoiding failure.

The goal is learning quickly.

High-performance organizations ask:

What did we learn?

What should we improve?

What should we stop doing?

What should we do differently next time?

This mindset transforms setbacks into assets.

Organizations become smarter with every experiment.

They accumulate knowledge while competitors remain stuck.

AI as a Learning Accelerator

Artificial intelligence offers organizations a remarkable opportunity to accelerate learning.

Employees can:

- Access information instantly.
- Explore new ideas quickly.

- Receive coaching and guidance.
- Generate alternatives.
- Analyze trends.
- Improve problem-solving capabilities.

However, AI only creates value when employees use it effectively.

This requires curiosity.

Training.

Experimentation.

Practice.

Organizations that encourage responsible AI exploration often discover opportunities that leadership alone would never identify.

Innovation becomes distributed throughout the organization.

Ideas emerge from every level.

Learning accelerates.

Performance improves.

Building a Culture of Ownership

Innovation thrives when employees feel ownership.

Ownership means:

"I care about the outcome."

"I take responsibility."

"I look for solutions."

"I contribute."

People rarely innovate because they are told to innovate.

They innovate because they care.

Leaders create ownership by:

- Sharing information openly.
- Explaining the why behind decisions.
- Involving employees in problem-solving.
- Recognizing contributions.
- Celebrating learning and improvement.

Ownership transforms employees from participants into partners.

And partners think differently.

The Leadership Role

Culture does not emerge by accident.

It is created intentionally.

And leadership shapes culture more than any other factor.

Employees watch leaders closely.

They observe:

What leaders reward.

What leaders tolerate.

What leaders prioritize.

What leaders discuss.

What leaders ignore.

If leaders talk about innovation but punish mistakes, employees notice.

If leaders promote learning but never invest in development, employees notice.

If leaders encourage experimentation but demand perfection, employees notice.

Culture follows behavior, not slogans.

Leaders must model the culture they want to create.

The Learning Organization Framework

Organizations seeking to thrive in the AI era should focus on five cultural commitments:

Curiosity

Encourage questions, exploration, and continuous improvement.

Courage

Support experimentation and intelligent risk-taking.

Collaboration

Break down silos and promote knowledge sharing.

Accountability

Own results and learn from outcomes.

Growth

Commit to continuous development for individuals and teams.

Together, these commitments create a culture capable of adapting continuously.

Leadership Reflection

Ask yourself:

- Do people feel safe sharing ideas?
- Do employees openly discuss mistakes?
- Is learning rewarded?
- Is innovation everyone's responsibility?
- Are leaders modeling curiosity and growth?
- Does the organization celebrate improvement as much as achievement?

The answers reveal the true state of the culture.

Not the culture written on posters.

The culture that actually exists.

Key Takeaways

1. Culture is the operating system of the organization.
2. Adaptability is becoming one of the most important competitive advantages.
3. Innovation should be everyone's responsibility, not the responsibility of a single department.
4. Continuous learning is essential in the AI era.
5. Psychological safety accelerates innovation and performance.
6. Organizations should reward learning, experimentation, and improvement.

7. AI can accelerate learning when employees are encouraged to explore and adapt.
8. Leaders shape culture through their actions more than their words.

The organizations that thrive in the age of AI will not simply have better technology.

They will have better cultures.

Cultures that learn faster.

Cultures that innovate continuously.

Cultures that embrace change rather than fear it.

But innovation and learning alone are not enough.

As AI becomes more deeply embedded into organizational life, another critical issue emerges:

Trust.

Without trust, adoption slows.

Without trust, resistance grows.

Without trust, transformation fails.

The next challenge for leaders is building trust in a world where intelligent systems increasingly influence decisions, recommendations, and outcomes.

Next Chapter: Chapter 8 , Building Trust in an Age of Intelligent Machines: Transparency, Ethics, and Responsible AI.

This chapter addresses one of the biggest concerns executives, employees, boards, and customers have about AI: trust, ethics, governance, transparency, and responsible deployment.

IN THE
AGE
OF
AI

CHAPTER 8

Building Trust in an Age of Intelligent Machines

Transparency, Ethics, and Responsible AI

Trust has always been the foundation of high-performance organizations.

Customers buy from organizations they trust.

Employees follow leaders they trust.

Investors support organizations they trust.

Communities welcome organizations they trust.

Without trust, relationships weaken.

Without trust, collaboration suffers.

Without trust, performance declines.

Artificial intelligence introduces a new challenge for leaders:

How do we maintain and strengthen trust when increasingly important decisions are influenced by intelligent systems?

The answer may determine whether AI becomes a powerful force for progress or a source of resistance and skepticism.

Technology can be purchased.

Trust must be earned.

The Trust Gap

Many organizations are moving faster with AI than their employees, customers, and stakeholders are comfortable with.

Leaders see opportunity.

Employees see uncertainty.

Customers see questions.

Boards see risk.

This creates what I call the Trust Gap.

The Trust Gap exists whenever technological adoption moves faster than understanding.

People tend to trust what they understand.

They tend to question what they do not understand.

If employees don't understand how AI is being used, they become cautious.

If customers don't understand how their data is being used, they become skeptical.

If leaders cannot explain how AI supports decision-making, confidence begins to erode.

Closing the Trust Gap requires transparency, communication, and responsible leadership.

Trust Is Built Before Technology Is Implemented

Many organizations wait until after implementation to address trust concerns.

This is a mistake.

Trust should be considered from the beginning.

Before launching an AI initiative, leaders should ask:

- How will this affect employees?
- How will this affect customers?
- What concerns might people have?
- How will we communicate our intentions?
- What safeguards should be established?

Organizations that proactively address these questions create stronger adoption and better outcomes.

Trust is not something that can be added later.

It must be designed into the process.

Transparency Creates Confidence

People do not expect leaders to have all the answers.

They do expect honesty.

Transparency means helping people understand:

Why AI is being implemented.

What it will be used for.

What it will not be used for.

How decisions are being made.

What protections are in place.

Transparency reduces fear because it reduces uncertainty.

Consider two scenarios.

In the first, employees discover new AI systems have been implemented without explanation.

Rumors begin circulating.

Concerns spread.

Resistance increases.

In the second, leaders explain the purpose, expected benefits, limitations, and safeguards before implementation begins.

Questions are encouraged.

Feedback is welcomed.

Concerns are addressed.

The technology remains the same.

The outcome is very different.

Trust grows when communication grows.

Responsible AI Begins with Leadership

Technology itself has no values.

People provide the values.

Artificial intelligence does not determine what is right or wrong.

Leaders do.

This is why responsible AI is fundamentally a leadership responsibility.

Every organization should establish clear principles that guide AI use.

Examples include:

- Respect for privacy
- Fairness

- Transparency
- Accountability
- Security
- Human oversight
- Responsible innovation

These principles provide guidance when difficult decisions arise.

Without principles, organizations risk making decisions based solely on efficiency or profitability.

High-performance organizations recognize that long-term trust is more valuable than short-term gains.

The Importance of Human Oversight

One of the most dangerous assumptions leaders can make is believing that AI is always correct.

It is not.

AI can make mistakes.

It can generate inaccurate information.

It can misunderstand context.

It can reinforce flawed assumptions.

It can produce recommendations that appear reasonable but are ultimately incorrect.

This is why human oversight remains essential.

Artificial intelligence should support decision-making.

It should not replace accountability.

At the end of the day, leaders remain responsible for organizational decisions.

The accountability cannot be delegated to an algorithm.

Organizations that understand this distinction are far less likely to encounter serious problems.

Fairness Matters

Trust requires fairness.

Employees want to know they will be treated fairly.

Customers want to know they will be treated fairly.

Communities want to know they will be treated fairly.

When organizations deploy AI systems, they must consider how those systems impact different groups of people.

Questions leaders should ask include:

- Are decisions being applied consistently?
- Could certain groups be disadvantaged?
- Are we monitoring outcomes?
- Do people have opportunities to appeal or challenge decisions?
- Are humans reviewing sensitive decisions?

These questions are not merely compliance issues.

They are trust issues.

Fairness strengthens credibility.

Credibility strengthens trust.

Trust strengthens performance.

Data Stewardship

Data is often called the fuel of AI.

That description is accurate.

However, many leaders overlook an important reality:

Data often represents people.

Customer information.

Employee information.

Partner information.

Community information.

Trust requires treating this information responsibly.

Organizations must be clear about:

- What data is collected
- Why it is collected
- How it is used
- Who has access to it
- How it is protected

People are generally willing to share information when they believe it will be used responsibly.

They become reluctant when they believe organizations are careless or deceptive.

Responsible data stewardship is no longer optional.

It is a core leadership responsibility.

The Ethics Question

As AI becomes more capable, organizations will increasingly face ethical decisions.

Should a process be automated?

Should a recommendation be followed?

Should a prediction influence a decision?

Should personal data be used in a particular way?

Technology may help evaluate options.

It cannot determine values.

Values remain human.

The organizations that thrive over the long term will establish ethical frameworks before they need them.

Waiting until a crisis occurs is often too late.

Ethics should guide innovation, not follow it.

Building an AI Governance System

High-performance organizations establish governance systems to ensure AI is used responsibly.

Governance does not mean bureaucracy.

It means clarity.

Effective AI governance typically addresses:

Purpose

Why are we using AI?

Accountability

Who is responsible for outcomes?

Oversight

Who reviews and monitors performance?

Risk Management

What could go wrong?

Compliance

Are we meeting legal and regulatory requirements?

Continuous Improvement

How will we learn and improve over time?

Governance provides structure.

Structure creates consistency.

Consistency creates trust.

Trust as a Competitive Advantage

Many leaders view trust as a soft concept.

In reality, trust is a strategic asset.

Organizations with high trust typically experience:

- Faster adoption of change
- Better collaboration
- Stronger customer loyalty
- Higher employee engagement
- Greater innovation
- Better decision-making

Trust reduces friction.

When trust is high, organizations move faster.

When trust is low, everything becomes harder.

Approvals take longer.

Communication becomes guarded.

Innovation slows.

Resistance increases.

Trust may be invisible on a balance sheet, but its impact is unmistakable.

The Leadership Challenge

The AI era presents leaders with a unique challenge.

Organizations must move quickly enough to remain competitive.

At the same time, they must move thoughtfully enough to maintain trust.

Too much caution can create stagnation.

Too much speed can create damage.

The most effective leaders find balance.

They embrace innovation.

They maintain accountability.

They encourage experimentation.

They establish safeguards.

They create progress without sacrificing trust.

This balancing act may become one of the defining leadership responsibilities of the coming decade.

Leadership Reflection

Ask yourself:

- Do employees understand how AI is being used?
- Have we clearly communicated our intentions?
- Do we have guiding principles for AI deployment?
- Are humans maintaining appropriate oversight?
- Do stakeholders trust our decision-making process?
- Are we moving quickly without compromising our values?

The answers reveal whether trust is being strengthened or weakened.

Key Takeaways

1. Trust is the foundation of successful AI adoption.
2. Transparency reduces fear and increases confidence.
3. Responsible AI begins with leadership, not technology.
4. Human oversight remains essential.
5. Fairness, privacy, and accountability must be built into AI systems.
6. Governance creates consistency and reduces risk.
7. Organizations that maintain trust move faster and perform better.
8. Trust is becoming a strategic competitive advantage in the AI era.

The organizations that successfully navigate the age of AI will not simply be known for technological innovation.

They will be known for responsible innovation.

They will create environments where employees, customers, and stakeholders trust both the technology and the leaders who deploy it.

And that trust begins with a resource that powers every successful AI initiative:

Data.

Without data, AI cannot learn.

Without quality data, AI cannot perform effectively.

Without a data strategy, even the most advanced AI systems struggle to create value.

The next step in building a high-performance AI-enabled organization is understanding how to treat data not merely as information, but as a strategic asset.

Next Chapter: Chapter 9 , Data as a Strategic Asset: The Foundation of Every Successful AI Initiative.

This chapter guides executives through their thinking about data strategically, without becoming technical, and how data quality, accessibility, governance, and decision-making create the foundation for AI success.

CHAPTER 9

Data as a Strategic Asset

The Foundation of Every Successful AI Initiative

Many leaders become fascinated by artificial intelligence while overlooking the resource that makes artificial intelligence possible.

Data.

Artificial intelligence cannot create value without it.

AI learns from data.

Analyzes data.

Identifies patterns within data.

Generates insights from data.

Makes predictions based on data.

Without data, AI is like a high-performance race car without fuel.

The technology may be impressive, but it cannot move the organization forward.

Unfortunately, many organizations discover this reality too late.

They invest heavily in AI platforms only to realize that their data is fragmented, inconsistent, inaccessible, outdated, or unreliable.

The result is frustration.

The technology works.

The foundation does not.

This chapter explores why data has become one of the most important strategic assets in the modern organization and how leaders can leverage it to create sustainable competitive advantage.

Data Is the New Infrastructure

For decades, organizations viewed infrastructure primarily in physical terms.

Buildings.

Equipment.

Machinery.

Transportation systems.

Technology has changed that perspective.

Today, data is infrastructure.

It powers decision-making.

It supports operations.

It improves customer experiences.

It enables innovation.

It fuels artificial intelligence.

Organizations that manage data effectively gain a significant advantage.

Organizations that neglect it often struggle regardless of how much technology they purchase.

The future belongs to organizations that treat data as a strategic asset rather than an administrative byproduct.

The Hidden Value of Data

Many organizations possess far more valuable data than they realize.

Customer interactions.

Sales histories.

Operational metrics.

Employee information.

Service records.

Market intelligence.

Financial transactions.

Performance indicators.

Every day, organizations generate enormous amounts of information.

The challenge is not collecting data.

The challenge is extracting value from it.

Imagine a company sitting on millions of dollars of inventory without realizing it.

That would seem absurd.

Yet organizations routinely sit on vast amounts of valuable data without fully utilizing it.

The organizations that learn how to transform information into insight create powerful advantages.

From Data to Decisions

Data by itself creates little value.

Information creates awareness.

Insight creates understanding.

Action creates results.

The journey looks like this:

Data → Information → Insight → Decision → Action → Results

Many organizations become trapped in the early stages.

They collect information endlessly.

They generate reports.

They build dashboards.

They create metrics.

Yet performance remains unchanged.

Why?

Because insight without action creates no value.

The purpose of data is not reporting.

The purpose of data is better decision-making.

Leaders should constantly ask:

What decisions should this information help us make?

That question shifts the conversation from measurement to performance.

Quality Matters More Than Quantity

One of the biggest misconceptions about AI is that more data automatically produces better results.

It does not.

Poor-quality data often creates poor-quality outcomes.

Consider a simple principle:

Bad data in.

Bad decisions out.

If customer information is inaccurate, customer insights become unreliable.

If operational data is incomplete, recommendations become flawed.

If performance metrics are inconsistent, leaders lose confidence in the results.

High-performance organizations focus on quality before quantity.

They emphasize:

- Accuracy
- Consistency
- Completeness
- Timeliness
- Reliability

Trustworthy data leads to trustworthy decisions.

Trustworthy decisions lead to better outcomes.

Breaking Down Data Silos

One of the most common obstacles organizations face is the existence of data silos.

Sales has data.

Marketing has data.

Finance has data.

Operations has data.

Human resources has data.

Each department possesses valuable information.

Yet the information often remains isolated.

The result is fragmented understanding.

Departments optimize locally while the organization struggles globally.

AI creates an opportunity to connect information across functions.

When leaders can see relationships across the entire organization, better decisions become possible.

Customer patterns emerge.

Operational inefficiencies become visible.

Growth opportunities become clearer.

The organizations that break down silos often uncover value that was hiding in plain sight.

Data Literacy: A Leadership Requirement

Not every leader needs to become a data scientist.

However, every leader must become data literate.

Data literacy means understanding:

- What information matters.
- How to interpret information.
- How to ask better questions.
- How to evaluate evidence.
- How to make informed decisions.

In the AI era, leaders who ignore data place themselves at a disadvantage.

The best leaders combine experience, judgment, and evidence.

They balance intuition with information.

They recognize patterns without becoming prisoners of assumptions.

Data literacy is becoming as important as financial literacy.

It is no longer optional.

Creating a Data-Driven Culture

Technology alone does not create a data-driven organization.

Culture does.

Employees must learn to value evidence.

Managers must learn to ask better questions.

Leaders must model fact-based decision-making.

A data-driven culture does not eliminate judgment.

It improves judgment.

The goal is not replacing human thinking.

The goal is supporting human thinking with better information.

Organizations create data-driven cultures when they:

- Share information openly.
- Encourage curiosity.
- Measure what matters.
- Focus on learning rather than blame.
- Use evidence to improve decisions.

The result is an organization that learns faster and performs better.

Data and Customer Understanding

One of the greatest opportunities created by data is deeper customer understanding.

Organizations can now learn:

- What customers value.
- How customers behave.
- What customers need.
- Why customers leave.

- What experiences drive loyalty.

Historically, organizations often relied on assumptions.

Today, leaders have access to far richer information.

The organizations that understand their customers best often outperform competitors.

Not because they possess better products.

Because they create better experiences.

Customer understanding becomes customer advantage.

Data and Innovation

Data is not only useful for operations.

It is also valuable for innovation.

Organizations can use information to identify:

- Emerging trends.
- Changing customer preferences.
- New market opportunities.
- Product improvements.
- Service enhancements.

Innovation becomes more informed.

Experiments become more targeted.

Investments become more strategic.

Rather than relying solely on intuition, organizations can combine creativity with evidence.

This combination often produces extraordinary results.

Data Governance: Protecting the Asset

Every valuable asset requires stewardship.

Data is no exception.

Organizations must establish clear practices for:

- Data quality.
- Data access.
- Privacy.
- Security.
- Compliance.
- Accountability.

Without governance, confusion emerges.

Without accountability, quality declines.

Without security, trust erodes.

Governance is not bureaucracy.

It is responsible stewardship.

The goal is ensuring that data remains accurate, accessible, secure, and valuable.

Organizations that treat data responsibly strengthen both performance and trust.

The AI Connection

Artificial intelligence magnifies the importance of data.

AI systems learn from available information.

The better the information, the better the outcomes.

Organizations that have invested in data quality often accelerate AI adoption more successfully.

Organizations with weak data foundations often struggle.

This is why leaders should view data strategy and AI strategy as inseparable.

One supports the other.

Neither succeeds independently.

Data is the foundation.

AI is the amplifier.

The Competitive Advantage

As AI becomes more widespread, technology itself will become increasingly accessible.

Competitive advantage will come from how effectively organizations use it.

That effectiveness depends largely on data.

Organizations that can:

- Collect meaningful information,
- Generate valuable insights,
- Make better decisions,
- Learn continuously,
- Act quickly,

will consistently outperform those that cannot.

The advantage will not come from possessing the most data.

It will come from creating the most value from the data available.

Leadership Reflection

Consider your organization.

- Do we treat data as a strategic asset?

- Is our information trustworthy?
- Are important insights accessible?
- Do departments share information effectively?
- Are leaders comfortable using data to support decisions?
- Does our data strategy support our AI strategy?

The answers reveal the strength of the foundation upon which future success will be built.

Key Takeaways

1. Data is the foundation of every successful AI initiative.
2. Organizations should treat data as a strategic asset rather than an administrative byproduct.
3. Data creates value only when it improves decision-making and action.
4. Quality matters more than quantity.
5. Breaking down data silos creates organizational insight.
6. Data literacy is becoming a core leadership competency.
7. Data-driven cultures learn and adapt faster.
8. Data strategy and AI strategy are inseparable.

The organizations that thrive in the age of AI will not simply possess powerful technology.

They will possess strong foundations.

They will understand their customers.

They will make better decisions.

They will learn faster than competitors.

And they will use data to create extraordinary value.

But technology, data, and organizational design alone are not enough.

Ultimately, transformation succeeds or fails because of leadership.

The next section of this book focuses on the single most important factor in building a high-performance AI-enabled organization:

The leader.

Next Chapter: Chapter 10 , The New Leadership Imperative: What Leaders Must Do Differently in the AI Era.

This chapter begins Part III and positions leadership, not technology, as the ultimate differentiator in high-performance organizations.

PART III

LEADING PEOPLE THROUGH TRANSFORMATION

IN THE
AGE
OF
AI

CHAPTER 10

The New Leadership Imperative

What Leaders Must Do Differently in the AI Era

Every era creates a new leadership challenge.

For one generation of leaders, the challenge was industrialization.

For another, it was globalization.

For others, it was digital transformation.

Today's leaders face a different challenge:

Leading organizations through the age of artificial intelligence.

The challenge is not simply understanding technology.

It is helping people, organizations, and cultures adapt to a world that is changing faster than ever before.

This distinction is important.

Technology may drive change.

Leadership determines whether organizations benefit from that change.

Throughout this book, we have explored artificial intelligence, organizational design, culture, trust, and data. Yet none of these elements reach their full potential without effective leadership.

Technology does not transform organizations.

People do.

And people look to leaders for direction.

The Leadership Shift

For decades, leadership often centered around expertise.

Leaders were expected to know more than others.

Have the answers.

Provide direction.

Make decisions.

Control outcomes.

While expertise remains valuable, the AI era is creating a different leadership reality.

No single leader can know everything.

The pace of change is too fast.

The volume of information is too large.

The complexity of decisions is too great.

Leaders must shift from being the person with all the answers to becoming the person who creates the environment where the best answers can emerge.

This requires humility.

Curiosity.

Collaboration.

Continuous learning.

The strongest leaders of the future will not necessarily be those who know the most.

They will be those who learn the fastest.

From Command and Control to Inspire and Enable

Many organizations were built around command-and-control leadership models.

Managers directed.

Employees executed.

Information flowed downward.

Decisions flowed upward.

This model worked reasonably well in stable environments.

The AI era is anything but stable.

Today's organizations need leaders who can inspire and enable rather than simply command and control.

Employees increasingly possess access to information, tools, and insights that were once available only to executives.

As a result, leadership becomes less about authority and more about influence.

Less about control and more about empowerment.

Less about directing every action and more about creating alignment.

The goal is not to remove accountability.

The goal is to unlock performance.

Clarity Becomes a Superpower

One of the greatest leadership responsibilities in the AI era is creating clarity.

When change accelerates, confusion often follows.

Employees hear conflicting messages.

Markets evolve.

Technologies emerge.

Priorities shift.

People begin asking:

What matters most?

What should I focus on?

How do I succeed?

What does the future look like?

Leaders who provide clarity create confidence.

Leaders who create confusion generate anxiety.

Clarity helps organizations move faster because people understand where they are going and why.

The ability to simplify complexity may become one of the most valuable leadership skills of the future.

Leaders Must Become Chief Learners

Historically, leaders were often viewed as teachers.

In the future, they must also become learners.

The organizations that thrive will be led by individuals who actively seek new knowledge, challenge assumptions, and remain open to change.

The moment leaders believe they have learned everything they need to know, growth begins to slow.

Employees take their cues from leadership.

When leaders demonstrate curiosity, others become curious.

When leaders embrace learning, others become more willing to learn.

When leaders experiment responsibly, others become more willing to innovate.

Learning is contagious.

So is stagnation.

Leaders choose which culture they create.

The Importance of Questions

Many leaders measure themselves by the quality of their answers.

High-performance leaders increasingly measure themselves by the quality of their questions.

Questions drive discovery.

Questions challenge assumptions.

Questions uncover opportunities.

Questions reveal risks.

Artificial intelligence can provide information.

Human beings must determine what questions are worth asking.

Consider the difference between these questions:

How can we reduce costs?

Versus:

How can we create more value?

How can we automate work?

Versus:

How can we improve human performance?

How do we implement AI?

Versus:

How do we become a higher-performing organization?

The questions leaders ask shape the conversations that follow.

The conversations shape decisions.

The decisions shape outcomes.

Building Trust During Uncertainty

Periods of transformation create uncertainty.

Uncertainty creates anxiety.

Anxiety creates resistance.

One of the leader's most important responsibilities is building trust during periods of uncertainty.

Employees do not expect leaders to predict the future perfectly.

They do expect honesty.

Transparency.

Consistency.

Authenticity.

Trust grows when leaders communicate openly.

When they acknowledge challenges.

When they explain decisions.

When they follow through on commitments.

When trust is present, organizations move faster.

When trust is absent, even simple changes become difficult.

Trust remains one of the highest-return investments a leader can make.

Developing People, Not Just Performance

Many leaders focus heavily on performance.

Performance matters.

Results matter.

Execution matters.

However, sustainable performance requires continuous development.

The AI era places a premium on learning, adaptability, and growth.

As a result, leaders must view talent development as a strategic responsibility.

The question is no longer:

How do we get more from our people?

The better question is:

How do we help our people become more capable?

Organizations that invest in capability development create long-term advantages.

Skills improve.

Confidence grows.

Adaptability increases.

Innovation accelerates.

The future belongs to organizations that develop people continuously.

Balancing Technology and Humanity

One of the greatest leadership challenges of the coming decade will be balancing technological advancement with human needs.

Leaders must ask:

How can technology improve performance?

How can it improve employee experiences?

How can it strengthen customer relationships?

How can it support innovation?

How can it reinforce organizational values?

Technology should serve people.

People should not become servants of technology.

The strongest leaders never lose sight of this principle.

They recognize that organizations exist to create value through people.

Technology is a powerful enabler.

It is not the purpose.

The Five Leadership Commitments

High-performance leaders in the AI era embrace five commitments.

Commitment 1: Lead with Purpose

Keep the mission at the center of every decision.

Commitment 2: Learn Continuously

Remain curious, adaptable, and open to new ideas.

Commitment 3: Develop People

Invest in capability, confidence, and growth.

Commitment 4: Build Trust

Create transparency, consistency, and credibility.

Commitment 5: Embrace Innovation

Encourage experimentation and responsible risk-taking.

Together, these commitments create organizations capable of thriving in a rapidly changing world.

Leadership Reflection

Ask yourself:

- Am I leading learning or protecting the status quo?
- Do employees understand where we are going?
- Am I developing people as aggressively as I pursue results?
- Am I creating clarity or contributing to confusion?

- Do people trust me during periods of uncertainty?
- Am I preparing the organization for the future or managing the present?

The answers reveal your readiness for the AI era.

The New Definition of Leadership

Perhaps the greatest leadership lesson of the AI age is this:

Leadership is no longer about having the most answers.

It is about creating the conditions where people, technology, and purpose come together to produce extraordinary results.

The leader of the future is not merely a manager of resources.

Not merely a decision-maker.

Not merely a strategist.

The leader of the future is an architect of possibility.

Someone who helps people see what can be achieved.

Someone who creates environments where learning flourishes.

Someone who turns uncertainty into opportunity.

Someone who helps organizations become better tomorrow than they are today.

That is the new leadership imperative.

Key Takeaways

1. Leadership remains the ultimate differentiator in the AI era.
2. Leaders must shift from being answer providers to learning facilitators.

3. Clarity is a critical leadership capability during periods of rapid change.
4. Trust enables organizations to move faster and adapt more effectively.
5. Leaders must focus on developing people as well as performance.
6. The future requires balancing technological advancement with human needs.
7. High-performance leaders embrace purpose, learning, development, trust, and innovation.
8. Leadership is ultimately about creating the conditions for success.

Technology may shape the future.

But leadership determines how successful organizations navigate it.

The next challenge is one every leader encounters during transformation:

Resistance.

Because regardless of how promising the future appears, people often struggle to let go of the familiar.

Understanding why resistance occurs, and how to address it effectively, may be one of the most valuable leadership skills of all.

Next Chapter: Chapter 11 , Managing Resistance and Fear: Helping People Navigate Change. This chapter provides leaders practical tools for overcoming resistance, building buy-in, and turning skeptics into participants during AI-driven transformation.

IN THE
AGE
OF
AI

CHAPTER 11

Managing Resistance and Fear

Helping People Navigate Change

Every leader wants change.

Until it affects them personally.

That statement may sound harsh, but it reflects a fundamental truth about human nature.

Most people support improvement.

Most people appreciate innovation.

Most people recognize the value of progress.

Yet when change begins affecting familiar routines, established skills, trusted relationships, and comfortable ways of working, resistance often emerges.

This is especially true when artificial intelligence enters the conversation.

Employees wonder:

Will my job change?

Will my skills still matter?

Will I be able to keep up?

Will I succeed in this new environment?

Managers wonder:

Will my role become less important?

How will I lead differently?

How will my team respond?

Executives wonder:

Will this transformation succeed?

Can we move quickly enough?

What risks are we overlooking?

Resistance is not unusual.

It is normal.

The organizations that successfully navigate AI transformation are not those that eliminate resistance.

They are those that understand it.

Understanding the Real Source of Resistance

Many leaders mistakenly believe people resist change.

In reality, people often resist loss.

They fear losing:

- Competence
- Confidence
- Status
- Relationships

- Control
- Predictability
- Security

When leaders view resistance through this lens, their approach changes dramatically.

Instead of asking:

"Why won't people support this?"

They begin asking:

"What are people worried about losing?"

That question creates empathy.

Empathy creates understanding.

Understanding creates better solutions.

The most effective leaders address the human concerns beneath the resistance.

The Emotional Journey of Change

Change is not merely an operational process.

It is an emotional process.

People often move through predictable stages when experiencing significant change.

Stage One: Awareness

People become aware that something is changing.

Questions emerge.

Speculation begins.

Stage Two: Concern

Individuals begin evaluating how the change might affect them personally.

Uncertainty grows.

Stage Three: Resistance

People express doubts, concerns, and objections.

This stage often receives the most attention.

Stage Four: Exploration

As understanding increases, people begin exploring possibilities.

Curiosity starts replacing fear.

Stage Five: Commitment

Individuals embrace the change and begin contributing to its success.

Understanding these stages helps leaders respond appropriately.

Resistance is often a temporary stage, not a permanent condition.

The Cost of Silence

One of the greatest mistakes leaders make during transformation is communicating too little.

When information is absent, people create their own explanations.

Rumors spread.

Assumptions multiply.

Fear fills the vacuum.

Consider the difference between these two approaches.

In the first organization, leaders quietly implement AI systems without explanation.

Employees hear fragments of information.

Speculation spreads.

Trust declines.

Resistance increases.

In the second organization, leaders communicate early and often.

They explain why changes are occurring.

They acknowledge concerns.

They answer questions.

They provide updates.

Trust grows.

Engagement increases.

The lesson is simple:

When in doubt, communicate.

Then communicate again.

The Four Questions Everyone Asks

Whenever significant change occurs, most people ask four questions.

Sometimes they ask them aloud.

More often, they ask them internally.

Question One:

Why is this happening?

People need context.

They need to understand the purpose.

Question Two:

What does this mean for me?

People naturally evaluate change through a personal lens.

Question Three:

Can I succeed?

People want confidence that they can adapt.

Question Four:

Will I be supported?

People need to know they will not face the challenge alone.

Leaders who answer these four questions consistently create significantly higher levels of engagement.

Fear Shrinks When Capability Grows

Many fears surrounding AI stem from uncertainty.

People fear what they do not understand.

They fear what they have not experienced.

They fear what they have not practiced.

The solution is not reassurance alone.

The solution is capability.

Training reduces fear.

Practice reduces fear.

Experience reduces fear.

Success reduces fear.

The more capable people become, the more confident they become.

This is why workforce development is such an important part of transformation.

Organizations that invest in helping employees succeed create far less resistance than organizations that simply demand adaptation.

The Difference Between Compliance and Commitment

Leaders often settle for compliance when they should be pursuing commitment.

Compliance means:

"I'll do it because I have to."

Commitment means:

"I'll do it because I believe in it."

Compliance creates minimal effort.

Commitment creates discretionary effort.

The difference is enormous.

Committed employees:

- Learn faster.
- Solve problems proactively.
- Share ideas.
- Help others.

- Support implementation.

Leaders create commitment by connecting change to purpose.

People want to be part of something meaningful.

When they understand how transformation benefits customers, employees, and the organization, commitment becomes more likely.

Turning Skeptics into Allies

Every transformation includes skeptics.

Some leaders view skeptics as obstacles.

High-performance leaders view them differently.

Skeptics often ask important questions.

They identify risks.

They challenge assumptions.

They expose blind spots.

The goal is not silencing skeptics.

The goal is engaging them.

Ask for their input.

Listen carefully.

Address legitimate concerns.

Involve them in solutions.

Many of the strongest advocates for change begin as skeptics who were treated respectfully and included in the process.

The Leadership Visibility Principle

During periods of transformation, leadership visibility matters.

People watch leaders closely.

They observe:

- What leaders say.
- What leaders do.
- What leaders prioritize.
- What leaders tolerate.

Employees want reassurance that leadership believes in the direction being taken.

They want confidence that leaders are engaged.

They want evidence that leaders understand the challenges.

Visible leadership creates confidence.

Absent leadership creates uncertainty.

Transformation cannot be delegated entirely.

Leaders must remain actively involved.

Celebrating Progress

One reason change initiatives fail is that organizations focus exclusively on the finish line.

Employees work hard.

Teams make progress.

Small victories occur.

Yet leadership discusses only what remains unfinished.

This can be discouraging.

Progress deserves recognition.

Every successful transformation consists of many small wins.

Learning a new skill.

Improving a process.

Successfully completing a pilot.

Achieving an adoption milestone.

Helping a customer more effectively.

Recognition reinforces momentum.

Momentum accelerates change.

Leaders who celebrate progress create energy.

The Change Leadership Framework

When helping people navigate AI-driven transformation, leaders should focus on five priorities.

Clarity

Explain what is changing and why.

Communication

Provide frequent, honest updates.

Capability

Invest in training and development.

Connection

Help people see how they fit into the future.

Confidence

Create opportunities for success and growth.

These five elements dramatically reduce resistance and increase engagement.

The Opportunity Hidden Inside Resistance

Many leaders view resistance as a problem.

Sometimes resistance is actually valuable information.

Resistance may reveal:

- Communication gaps.
- Training needs.
- Process weaknesses.
- Trust issues.
- Legitimate concerns.

Rather than fighting resistance, effective leaders learn from it.

They ask:

What is this resistance trying to teach us?

That question often uncovers opportunities for improvement.

Leadership Reflection

Consider a transformation currently occurring in your organization.

Ask yourself:

- Have we explained why the change matters?
- Do employees understand how it affects them?

- Have we invested enough in capability development?
- Are leaders visible and engaged?
- Are we listening as much as we are talking?
- Have we celebrated progress along the way?

The answers reveal how effectively the organization is managing change.

Key Takeaways

1. People often resist perceived loss more than they resist change itself.
2. Transformation is an emotional process as well as an operational process.
3. Communication is one of the most powerful tools for reducing resistance.
4. Capability development reduces fear and increases confidence.
5. Commitment creates better outcomes than compliance.
6. Skeptics can become valuable allies when engaged properly.
7. Leadership visibility is critical during periods of uncertainty.
8. Resistance often contains valuable information that can improve implementation.

The organizations that succeed in the AI era will not be those that avoid resistance.

They will be those that help people move through it successfully.

And one of the most effective ways to reduce resistance is to prepare people for the future.

That preparation begins with developing the skills, knowledge, and capabilities necessary to thrive in an AI-enabled world.

The next challenge for leaders is not simply implementing technology.

It is helping their workforce become ready for what comes next.

Next Chapter: Chapter 12 , Upskilling the Workforce: Preparing Employees for an AI-Augmented Future.

This chapter provides a practical roadmap for helping employees develop the skills, mindset, and confidence needed to work effectively alongside AI.

IN THE
AGE
OF
AI

CHAPTER 12

Upskilling the Workforce

Preparing Employees for an AI-Augmented Future

Technology has never been the greatest asset of an organization.

People have.

Buildings depreciate.

Equipment becomes obsolete.

Software gets replaced.

Technology evolves.

People learn.

People adapt.

People innovate.

People create value.

That reality remains true in the age of artificial intelligence.

In fact, as AI becomes more capable, the value of a skilled, adaptable, and engaged workforce increases.

The organizations that thrive will not necessarily be those with the most advanced technology.

They will be those with the most capable people.

This presents leaders with a critical challenge:

How do we prepare employees for a future where artificial intelligence becomes part of everyday work?

The answer begins by understanding a simple truth.

The future is not about replacing people with AI.

The future is about helping people work more effectively with AI.

The Workforce Transformation Opportunity

Throughout history, technological advances have changed the nature of work.

Agricultural innovations changed farming.

Industrial innovations changed manufacturing.

Computers changed office work.

The internet changed communication and commerce.

Artificial intelligence is creating another transformation.

Tasks will change.

Workflows will change.

Skills will change.

Expectations will change.

But change does not automatically mean displacement.

For many organizations, the greatest opportunity is not workforce reduction.

It is workforce elevation.

Employees can spend less time on routine activities and more time on high-value work.

Less time searching for information.

More time solving problems.

Less time producing reports.

More time creating strategy.

Less time performing repetitive tasks.

More time serving customers and innovating.

This shift can significantly increase organizational performance.

The Skills That Matter Most

Many leaders ask:

"What skills should our employees develop?"

The answer may surprise them.

While technical knowledge is important, the most valuable future skills are often deeply human.

Critical Thinking

Employees must learn how to evaluate information, question assumptions, and make sound decisions.

AI can provide recommendations.

Humans must determine whether those recommendations make sense.

Problem Solving

Organizations increasingly need people who can identify challenges and create solutions.

Technology supports this process.

People drive it.

Communication

As information becomes more abundant, clarity becomes more valuable.

Employees who communicate effectively create stronger teams, better customer experiences, and improved outcomes.

Collaboration

Future work will involve greater collaboration between people, teams, and intelligent systems.

The ability to work well with others becomes increasingly important.

Adaptability

The willingness to learn and adjust may become the most important skill of all.

Technology changes.

Markets change.

Organizations change.

Employees who adapt thrive.

Leadership

Leadership is no longer reserved for individuals with management titles.

Every employee influences others.

Every employee contributes to culture.

Every employee can demonstrate leadership.

Organizations should develop leadership capability at every level.

Developing AI Literacy

Just as organizations once focused on computer literacy, they must now focus on AI literacy.

AI literacy does not mean everyone becomes a programmer or data scientist.

It means employees understand:

- What AI is.
- What AI can do.
- What AI cannot do.
- How to use AI responsibly.
- How to evaluate AI-generated information.
- How AI supports their work.

When employees understand AI, fear decreases.

Confidence grows.

Adoption improves.

Innovation accelerates.

Knowledge creates opportunity.

The Learning Mindset

Perhaps the most important workforce capability is not a skill.

It is a mindset.

The mindset of continuous learning.

Employees who believe learning ends after formal education often struggle during periods of rapid change.

Employees who view learning as a lifelong responsibility tend to thrive.

High-performance organizations encourage employees to ask:

What do I need to learn next?

How can I improve?

What new capability would make me more valuable?

How can I contribute at a higher level?

This mindset creates resilience.

It creates adaptability.

It creates future readiness.

Moving Beyond Training Events

Many organizations still approach development as an event.

Employees attend a workshop.

Complete a course.

Participate in a seminar.

Then return to work.

While formal training remains valuable, it is no longer enough.

The future belongs to organizations that integrate learning into daily work.

Learning should occur:

- During projects.

- Through coaching.
- Through mentoring.
- Through experimentation.
- Through collaboration.
- Through AI-assisted learning tools.

Development becomes continuous rather than occasional.

The organization itself becomes a learning system.

The Role of Leaders in Workforce Development

Some leaders view development as the responsibility of human resources.

This perspective is incomplete.

Human resources can support development.

Leaders drive it.

Employees learn what leaders prioritize.

If leaders invest time in coaching, development accelerates.

If leaders discuss growth, learning becomes important.

If leaders reward improvement, capability expands.

The strongest leaders understand that one of their primary responsibilities is helping people become more capable.

Their success is measured not only by results.

It is measured by the growth of the people around them.

The Capability Multiplier

Artificial intelligence creates a remarkable opportunity.

Employees can access coaching, information, guidance, and insights instantly.

They can learn faster.

Practice faster.

Improve faster.

The combination of human ambition and AI support can create extraordinary outcomes.

Imagine an employee who:

- Receives instant feedback.
- Learns new skills rapidly.
- Accesses expertise on demand.
- Solves problems more effectively.
- Develops greater confidence.

That employee becomes significantly more valuable.

Now imagine an entire organization filled with employees who are continuously learning and improving.

That organization gains a powerful competitive advantage.

Creating a Culture of Growth

Upskilling succeeds when growth becomes part of the culture.

Employees should feel encouraged to:

- Learn.

- Experiment.
- Ask questions.
- Develop new capabilities.
- Share knowledge.

Organizations that punish mistakes often discourage learning.

Organizations that support growth accelerate learning.

The goal is not perfection.

The goal is continuous improvement.

Growth cultures consistently outperform fixed cultures because they adapt more effectively to change.

Investing in Potential

One of the most important leadership decisions involves where to invest resources.

Many organizations invest heavily in technology while underinvesting in people.

This creates imbalance.

Technology without capable people rarely delivers its full value.

The strongest organizations invest in both.

They view workforce development not as a cost but as a strategic investment.

They recognize that people create value.

Technology amplifies that value.

The combination produces extraordinary results.

Preparing for Jobs That Don't Yet Exist

One of the challenges of the AI era is that many future roles have not yet been defined.

Leaders cannot train employees exclusively for specific jobs.

They must prepare them for adaptability.

The focus shifts from teaching people exactly what to do toward helping them learn how to learn.

The most valuable employees of the future may not be those who possess the most knowledge today.

They may be those who can acquire new knowledge fastest tomorrow.

This changes how organizations think about development.

The goal becomes building learning agility.

The Workforce Advantage

As AI becomes more accessible, technology advantages may become increasingly temporary.

Workforce advantages can be more durable.

Organizations with capable, adaptable, engaged employees consistently outperform those that rely solely on technology.

Why?

Because people create innovation.

People build relationships.

People solve problems.

People exercise judgment.

People create culture.

People lead.

Technology enhances these capabilities.

It does not replace them.

Leadership Reflection

Ask yourself:

- Is our workforce prepared for the future?
- Are employees developing AI literacy?
- Do we reward learning and growth?
- Are leaders actively coaching and developing people?
- Do employees feel confident about their ability to adapt?
- Are we investing in capability development as aggressively as we invest in technology?

The answers reveal whether the organization is building a workforce advantage or merely implementing tools.

Key Takeaways

1. People remain the most important asset in high-performance organizations.
2. The future is about augmentation, not simply automation.
3. Critical thinking, communication, collaboration, adaptability, and leadership are becoming increasingly valuable.
4. AI literacy should become a core workforce competency.
5. Continuous learning is replacing episodic training.

6. Leaders play a critical role in workforce development.
7. Organizations should invest in capability development alongside technology investments.
8. Workforce adaptability may become one of the most important competitive advantages of the AI era.

The organizations that win in the age of AI will not merely deploy technology successfully.

They will develop people successfully.

And when capable people are equipped with intelligent tools, something powerful happens:

Teams begin performing at levels that neither humans nor machines can achieve independently.

The next step is learning how to build and lead those teams effectively.

Next Chapter: Chapter 13 , Building Human-AI Teams: Combining Human Judgment with Machine Intelligence.

This chapter strategically outlines how leaders can redesign work, workflows, and collaboration so that humans and AI complement one another rather than compete with one another.

PART III

LEADING PEOPLE THROUGH TRANSFORMATION

IN THE
AGE
OF
AI

CHAPTER 13

Building Human-AI Teams

Combining Human Judgment with Machine Intelligence

For years, discussions about artificial intelligence have often been framed as a competition.

Humans versus machines.

People versus technology.

Employees versus automation.

The reality is far more interesting.

The greatest opportunity is not competition.

It is collaboration.

The organizations creating the most value with AI are not asking:

"How do we replace people?"

They are asking:

"How do we help people perform at levels that were previously impossible?"

This question changes everything.

It shifts the focus from substitution to augmentation.

From elimination to enhancement.

From fear to opportunity.

The future belongs to organizations that learn how to combine human judgment with machine intelligence in ways that maximize the strengths of both.

Understanding Complementary Strengths

Humans and AI excel at different things.

Artificial intelligence is exceptionally good at:

- Processing large amounts of information.
- Identifying patterns.
- Automating repetitive tasks.
- Retrieving knowledge quickly.
- Generating options.
- Performing calculations.
- Scaling routine activities.

Human beings excel at:

- Leadership.
- Judgment.
- Creativity.
- Empathy.
- Ethics.
- Relationship-building.
- Contextual understanding.
- Strategic thinking.

Neither is superior in every situation.

Each brings unique strengths.

The goal is not deciding which is more valuable.

The goal is determining how to combine them effectively.

The organizations that accomplish this create a significant performance advantage.

The Calculator Lesson

Consider what happened when calculators became widely available.

Mathematicians did not disappear.

Accountants did not disappear.

Engineers did not disappear.

Instead, they became more productive.

The calculator handled routine calculations.

People focused on solving more complex problems.

Artificial intelligence is creating a similar opportunity across a much broader range of work.

AI handles many routine cognitive tasks.

Humans focus on judgment, innovation, leadership, and value creation.

The result is often higher performance from both.

The Human-AI Workflow

One of the simplest ways to understand human-AI collaboration is through workflow design.

Imagine a project requiring research, analysis, decision-making, communication, and implementation.

Traditionally, people performed every step manually.

In an AI-enabled workflow:

AI can gather information.

AI can summarize findings.

AI can identify patterns.

AI can generate alternatives.

Humans evaluate options.

Humans exercise judgment.

Humans make decisions.

Humans build support.

Humans execute and lead.

The work becomes faster without sacrificing quality.

The technology supports the people.

The people direct the technology.

This partnership often produces better outcomes than either could achieve independently.

Moving from Efficiency to Effectiveness

Many organizations initially adopt AI to improve efficiency.

Efficiency matters.

However, the larger opportunity is effectiveness.

Efficiency asks:

"How can we do things faster?"

Effectiveness asks:

"How can we achieve better outcomes?"

Consider a sales team.

AI might help representatives prepare for meetings more quickly.

That improves efficiency.

But the larger benefit may be that representatives spend more time building relationships and understanding customer needs.

That improves effectiveness.

The same principle applies across the organization.

AI creates time.

Leaders determine how that time is used.

High-performance organizations reinvest time into higher-value activities.

The Four Levels of Human-AI Collaboration

Organizations typically evolve through four stages.

Level One: Automation

AI performs repetitive tasks.

Examples include scheduling, reporting, and routine administration.

Level Two: Assistance

AI supports employees by providing information, recommendations, and content.

Examples include drafting reports, summarizing meetings, and answering questions.

Level Three: Augmentation

Humans and AI collaborate to improve decision-making, creativity, and performance.

The combined output exceeds what either could achieve independently.

Level Four: Transformation

Organizations redesign workflows, roles, and business models around human-AI collaboration.

New capabilities emerge.

New opportunities become possible.

The greatest value often occurs at Levels Three and Four.

Redesigning Roles

As AI becomes integrated into daily work, job roles will evolve.

This does not necessarily mean fewer people.

It often means different responsibilities.

Employees may spend:

Less time searching.

More time analyzing.

Less time reporting.

More time deciding.

Less time processing.

More time innovating.

Less time gathering information.

More time building relationships.

The nature of work shifts upward.

Employees focus increasingly on activities that require uniquely human capabilities.

This evolution can create greater engagement and greater value simultaneously.

The Leadership Role in Human-AI Teams

Leaders play a critical role in making human-AI collaboration successful.

They must establish:

Clear expectations.

Appropriate boundaries.

Responsible use guidelines.

Training and support.

Accountability.

Leaders also help employees understand that AI is a tool for enhancement rather than replacement.

The way leaders frame AI significantly influences adoption.

If AI is introduced as a threat, resistance grows.

If AI is introduced as a performance partner, curiosity increases.

Leadership shapes perception.

Perception shapes behavior.

Behavior shapes outcomes.

Maintaining Human Judgment

One of the greatest risks in the AI era is overreliance on technology.

As AI becomes more capable, people may become tempted to accept recommendations without sufficient evaluation.

This can be dangerous.

Technology should inform judgment.

Not replace it.

Leaders should encourage employees to ask:

Does this recommendation make sense?

What assumptions are being made?

What context may be missing?

What unintended consequences could occur?

Critical thinking becomes even more important when information is abundant.

Human judgment remains the final safeguard.

Building Confidence Through Experience

Many employees initially feel uncertain about working with AI.

The solution is experience.

People gain confidence by:

Using the tools.

Practicing new workflows.

Experimenting safely.

Learning from results.

The more employees experience success, the more comfortable they become.

Organizations should create opportunities for low-risk experimentation.

Small wins build confidence.

Confidence accelerates adoption.

Adoption creates value.

Measuring Team Performance

Traditional performance measures may need adjustment.

Leaders should evaluate not only individual productivity but also the effectiveness of human-AI collaboration.

Questions include:

- Are decisions improving?

- Is customer value increasing?
- Is innovation accelerating?
- Are employees becoming more productive?
- Are workflows becoming more effective?
- Is quality improving?

The goal is not measuring technology usage.

The goal is measuring outcomes.

Technology is only valuable when it contributes to results.

The New Productivity Equation

Historically, productivity often depended on:

Skill × Effort

In the AI era, the equation expands.

Performance increasingly becomes:

Skill × Effort × Technology

Two employees with similar skills may achieve dramatically different results depending on how effectively they leverage AI.

This creates a new leadership challenge.

Organizations must help employees develop both human capabilities and technological capabilities.

The combination becomes the differentiator.

The Human Advantage

As AI continues advancing, some leaders worry that human contribution will diminish.

The opposite may occur.

As technology handles more routine work, human capabilities become increasingly important.

Creativity becomes more valuable.

Judgment becomes more valuable.

Empathy becomes more valuable.

Leadership becomes more valuable.

Trust becomes more valuable.

Innovation becomes more valuable.

The organizations that recognize this dynamic will invest aggressively in both technology and people.

The organizations that focus exclusively on technology may miss the larger opportunity.

Leadership Reflection

Ask yourself:

- How effectively are humans and AI collaborating in our organization?
- Which tasks should be automated?
- Which tasks should remain human-centered?
- Are employees developing confidence in using AI?
- Are we redesigning work or merely adding tools?
- Are we measuring outcomes or simply technology adoption?

The answers reveal whether the organization is creating true augmentation or merely implementing software.

Key Takeaways

1. The future is based on collaboration between humans and AI, not competition.
2. Humans and AI possess complementary strengths.
3. The greatest value often comes from augmentation rather than automation.

4. Human judgment remains essential.
5. Organizations should redesign workflows, not simply deploy tools.
6. Leadership plays a critical role in adoption and success.
7. Confidence grows through experience and experimentation.
8. The most successful organizations invest in both people and technology.

The age of AI is creating extraordinary opportunities for performance improvement.

But performance alone is not enough.

Organizations must also help people understand where they are going, why they are going there, and how they can contribute.

That requires communication.

And in times of transformation, communication becomes one of the most important leadership tools available.

The next chapter explores how leaders create clarity, confidence, and commitment through effective communication during periods of rapid change.

Next Chapter: Chapter 14 , Communication During Transformation: How Great Leaders Create Clarity, Confidence, and Commitment. This chapter completes Part III and provides practical communication frameworks for leading AI-driven change successfully.

IN THE
AGE
OF
AI

CHAPTER 14

Communication During Transformation

How Great Leaders Create Clarity, Confidence, and Commitment

When organizations undergo transformation, most leaders focus on strategy.

They focus on technology.

They focus on budgets.

They focus on timelines.

They focus on implementation plans.

All of these are important.

Yet one factor often determines whether transformation succeeds or fails:

Communication.

Organizations rarely fail because people cannot handle change.

More often, they fail because people do not understand the change.

They do not understand:

Why it is happening.

What it means.

What is expected.

How they can contribute.

Where the organization is going.

In the absence of understanding, confusion grows.

When confusion grows, resistance follows.

The organizations that navigate transformation most successfully are often the organizations that communicate most effectively.

Communication Is a Leadership Responsibility

Many leaders assume communication is the responsibility of marketing, public relations, or human resources.

Those functions can provide valuable support.

However, communication during transformation is fundamentally a leadership responsibility.

Employees want to hear from leaders.

Customers want to hear from leaders.

Stakeholders want to hear from leaders.

People look to leadership for direction during periods of uncertainty.

Silence creates anxiety.

Visibility creates confidence.

The most effective leaders communicate early, often, and consistently.

Why Communication Breaks Down

Most leaders believe they communicate more than they actually do.

They announce a change.

Conduct a meeting.

Send an email.

Provide an update.

Then assume everyone understands.

Unfortunately, communication does not work that way.

People hear information through the filter of their own experiences, concerns, assumptions, and priorities.

What is perfectly clear to leadership may be completely unclear to employees.

This is why communication is not about what is said.

It is about what is understood.

High-performance leaders recognize this distinction.

They focus not only on transmitting information but also on ensuring understanding.

The Communication Gap

One of the greatest risks during transformation is the Communication Gap.

This occurs when leadership believes information has been communicated, but employees have not fully understood it.

The result can be costly.

Employees begin filling in the blanks.

Rumors emerge.

Misinformation spreads.

Anxiety increases.

Productivity declines.

Trust weakens.

One of the most effective practices leaders can implement is what I call the Information Check.

An Information Check occurs when leaders intentionally verify that communication has been received and understood correctly.

Rather than asking:

"Did everyone get the message?"

Leaders ask:

"What did you hear?"

"How do you understand this change?"

"What concerns do you have?"

"What questions remain?"

This simple practice often reveals significant gaps between what leaders intended to communicate and what employees actually understood.

The Whisper Line Lesson

There is a classic exercise often used in leadership and communication workshops.

Twenty people stand in a line.

A message is whispered to the first person.

That individual whispers the message to the second.

The second whispers it to the third.

The process continues until the final person receives the message.

When the last person shares what they heard, the message is often dramatically different from the original.

The lesson is obvious.

Information changes as it moves through people.

Organizations experience the same phenomenon every day.

The larger the organization, the greater the risk.

Leaders who conduct Information Checks reduce distortion and increase alignment.

In the AI era, where change occurs rapidly, this practice becomes even more important.

The Four Messages Every Transformation Requires

Employees do not need every detail immediately.

They do need four essential messages.

Message One: Why

Why are we making this change?

People support what they understand.

Purpose creates engagement.

Message Two: Vision

What future are we trying to create?

Vision creates direction.

People need to know where the organization is going.

Message Three: Expectations

What is expected of me?

Clarity reduces uncertainty.

Employees perform better when expectations are clear.

Message Four: Support

How will the organization help me succeed?

Support creates confidence.

People become more willing to embrace change when they know they will not face it alone.

Leaders who consistently communicate these four messages create stronger alignment throughout the organization.

The Importance of Repetition

One of the most misunderstood aspects of leadership communication is repetition.

Leaders often become tired of a message long before employees fully absorb it.

What feels repetitive to leadership often feels insufficient to others.

Great leaders repeat important messages consistently.

Not because people are incapable of understanding.

Because transformation requires reinforcement.

The mission should be repeated.

The vision should be repeated.

The priorities should be repeated.

The reasons for change should be repeated.

Repetition creates alignment.

Alignment accelerates execution.

Communicating with Honesty

Transformation creates uncertainty.

Employees know this.

Customers know this.

Stakeholders know this.

Leaders damage trust when they pretend uncertainty does not exist.

People do not expect perfection.

They do expect honesty.

Strong leaders say:

"Here's what we know."

"Here's what we don't know."

"Here's what we're learning."

"Here's what we're doing next."

This approach builds credibility.

Credibility strengthens trust.

Trust increases commitment.

Storytelling Creates Understanding

Facts inform.

Stories inspire.

Throughout history, leaders have used stories to help people understand change.

Stories create emotional connection.

Stories make ideas memorable.

Stories help people visualize possibilities.

When discussing AI transformation, leaders should share stories about:

- Customers benefiting from improvements.
- Employees learning new skills.
- Teams solving problems differently.
- Organizations creating new opportunities.

Stories transform abstract concepts into relatable experiences.

People remember stories long after they forget statistics.

Listening Is Communication

Many leaders view communication as speaking.

In reality, communication also involves listening.

Listening reveals:

- Concerns.
- Questions.
- Misunderstandings.
- Opportunities.
- Ideas.

The most effective communicators spend significant time listening.

They seek feedback.

They encourage questions.

They invite participation.

They treat communication as a dialogue rather than a broadcast.

Transformation succeeds faster when people feel heard.

The Visibility Principle

During transformation, visibility matters.

Employees want to see leaders.

Hear leaders.

Interact with leaders.

Ask questions.

Observe commitment.

Visible leaders create confidence.

Invisible leaders create uncertainty.

This does not require constant meetings.

It requires intentional presence.

Town halls.

Small group discussions.

Q&A sessions.

Site visits.

Informal conversations.

These interactions build trust and reinforce alignment.

Creating Commitment

The ultimate goal of communication is not awareness.

It is commitment.

Awareness means people know what is happening.

Commitment means people actively support it.

Commitment emerges when people:

Understand the purpose.

Believe in the vision.

Trust the leadership.

Feel supported.

See a role for themselves.

Great communication creates all five conditions.

The Communication Framework for AI Transformation

When leading AI-related change, every communication should answer five questions:

Why are we doing this?

Purpose.

What are we trying to achieve?

Vision.

What does this mean for you?

Personal relevance.

How will we help you succeed?

Support.

What happens next?

Action.

If leaders consistently address these questions, uncertainty decreases and engagement increases.

Leadership Reflection

Ask yourself:

- Are we communicating enough?
- Are employees truly understanding the message?
- Do we conduct Information Checks?
- Are leaders visible and accessible?
- Are we listening as much as we are speaking?
- Are we creating awareness or commitment?

The answers reveal the effectiveness of the organization's communication strategy.

Key Takeaways

1. Communication is one of the most important leadership responsibilities during transformation.
2. Communication success is measured by understanding, not transmission.
3. Information Checks help leaders verify understanding and alignment.
4. Repetition strengthens clarity and commitment.
5. Honesty builds trust during uncertainty.

6. Storytelling helps people understand and embrace change.
7. Listening is as important as speaking.
8. Effective communication creates commitment, not merely awareness.

The organizations that succeed in the AI era will communicate differently.

They will create clarity amid complexity.

Confidence amid uncertainty.

Commitment amid change.

They will understand that technology may accelerate transformation, but communication determines whether people choose to participate.

And when people understand the vision, trust the leadership, and embrace the opportunity, organizations become capable of extraordinary performance.

The next section of this book moves from leadership and culture into execution, exploring how AI can be used to create measurable improvements in productivity, decision-making, innovation, customer experience, and growth.

Next Chapter: Chapter 15 , AI and Operational Excellence: Improving Productivity, Quality, and Speed.

This chapter begins Part IV and shows leaders how AI creates tangible business value through improved execution and operational performance.

PART IV
USING AI TO CREATE EXTRAORDINARY PERFORMANCE

IN THE
AGE
OF
AI

CHAPTER 15

AI and Operational Excellence

Improving Productivity, Quality, and Speed

At some point, every leader asks the same question:

"What results will artificial intelligence actually produce for our organization?"

It is a fair question.

After all, organizations do not invest in technology for the sake of technology.

They invest to improve performance.

They invest to create value.

They invest to achieve outcomes that would otherwise be difficult, expensive, or impossible to achieve.

This is where operational excellence enters the conversation.

For decades, organizations have pursued operational excellence through better processes, stronger leadership, improved systems, workforce development, and continuous improvement.

Artificial intelligence does not replace these disciplines.

It enhances them.

The organizations achieving the greatest success with AI are using it to improve three fundamental drivers of performance:

Productivity. Quality. Speed.

When these three improve simultaneously, extraordinary results become possible.

The Productivity Opportunity

Productivity is often misunderstood.

Many people define productivity as working harder.

In reality, productivity is about creating more value with available resources.

Working harder is one option.

Working smarter is usually a better one.

Artificial intelligence creates opportunities to increase productivity by reducing the amount of time spent on low-value activities.

Consider how much time employees spend:

- Searching for information.
- Creating reports.
- Scheduling meetings.
- Summarizing documents.
- Answering repetitive questions.
- Updating records.
- Gathering data.
- Preparing presentations.

These activities are necessary.

However, they rarely represent the highest-value contribution employees can make.

AI can often perform or assist with many of these tasks.

As a result, employees gain time.

The question then becomes:

How will that time be used?

High-performance organizations reinvest the time into activities that create greater value.

Customer relationships.

Problem-solving.

Innovation.

Strategic thinking.

Leadership.

The goal is not simply doing more work.

The goal is doing more meaningful work.

Eliminating Friction

Every organization contains friction.

Approvals that take too long.

Information that is difficult to locate.

Processes that require unnecessary steps.

Communication delays.

Duplicated effort.

Manual tasks.

Friction slows performance.

Artificial intelligence can help identify and reduce many forms of organizational friction.

For example:

Employees can access information faster.

Workflows can become more streamlined.

Routine approvals can be automated.

Questions can be answered immediately.

Reports can be generated automatically.

When friction decreases, productivity increases naturally.

Organizations move faster because less energy is wasted overcoming obstacles.

Improving Quality

Operational excellence is not only about speed.

It is also about quality.

Fast mistakes are still mistakes.

High-performance organizations pursue both.

Artificial intelligence can improve quality in several ways.

It can:

- Detect errors.
- Identify inconsistencies.
- Monitor performance.

- Highlight risks.
- Support compliance.
- Improve accuracy.

Consider customer service.

AI can help representatives access information more quickly and consistently.

This can improve response quality.

Consider manufacturing.

AI can identify quality issues before products reach customers.

Consider finance.

AI can detect unusual transactions that may require attention.

In each case, the technology helps organizations reduce errors and improve outcomes.

The result is better performance and greater trust.

Accelerating Speed

Speed has become a strategic advantage.

Customers expect faster responses.

Markets change rapidly.

Competitors innovate continuously.

Organizations that move slowly often struggle.

Artificial intelligence helps accelerate:

- Information gathering.
- Analysis.

- Reporting.
- Customer interactions.
- Workflow execution.
- Decision preparation.

The key phrase is decision preparation.

AI often helps leaders arrive at decisions faster by organizing information and highlighting relevant insights.

This allows organizations to respond more quickly without sacrificing quality.

The goal is not rushing.

The goal is reducing unnecessary delays.

The Capacity Multiplier

One of the most significant benefits of AI is its ability to expand organizational capacity.

Traditionally, organizations increased capacity by hiring additional employees.

More work required more people.

AI creates another option.

Organizations can increase capacity by helping existing employees become more productive.

This creates a multiplier effect.

A manager can supervise more effectively.

A salesperson can engage more customers.

A customer service representative can resolve more issues.

A project manager can coordinate more initiatives.

The organization gains capacity without a proportional increase in resources.

This creates opportunities for growth, profitability, and innovation.

Operational Excellence Is Still Human

One of the biggest misconceptions about AI is that operational excellence becomes a technology issue.

It does not.

Operational excellence remains a leadership issue.

Technology can improve processes.

People improve performance.

Technology can provide information.

People create accountability.

Technology can automate tasks.

People create culture.

Organizations that achieve extraordinary operational performance recognize this distinction.

They use AI as an enabler.

They continue investing in leadership, culture, and capability development.

The Continuous Improvement Advantage

Many organizations view improvement as a periodic activity.

They conduct annual reviews.

Quarterly assessments.

Occasional process improvement projects.

The AI era creates opportunities for continuous improvement.

Real-time information becomes available.

Performance can be monitored continuously.

Patterns become visible more quickly.

Problems can be identified earlier.

This allows organizations to improve more frequently.

Small improvements accumulate.

Small improvements compound.

Over time, they create significant competitive advantages.

The organizations that learn and improve continuously often outperform organizations that rely on occasional breakthroughs.

Measuring What Matters

One danger organizations face is becoming overly focused on activity metrics.

Number of reports generated.

Number of automated tasks.

Number of AI tools deployed.

These metrics may be interesting.

They do not necessarily indicate success.

The true measures of success include:

- Productivity improvement.

- Quality improvement.
- Customer satisfaction.
- Employee engagement.
- Revenue growth.
- Cost reduction.
- Cycle-time reduction.
- Innovation outcomes.

Leaders should focus on outcomes rather than technology adoption alone.

Technology is a means.

Performance is the goal.

Operational Excellence and Employee Experience

Operational excellence benefits employees as well as organizations.

Employees often become frustrated by:

- Repetitive work.
- Administrative burdens.
- Inefficient systems.
- Poor communication.
- Time-consuming processes.

AI can reduce many of these frustrations.

Employees gain more time to focus on meaningful work.

This can improve:

- Job satisfaction.
- Engagement.
- Learning.
- Creativity.
- Performance.

Operational excellence and employee experience are not competing priorities.

When designed properly, they reinforce one another.

The Leader's Role

Leaders play a critical role in operational improvement.

They determine:

Where attention is focused.

Which problems are prioritized.

Which processes are redesigned.

Which outcomes matter most.

Technology can provide opportunities.

Leadership determines whether those opportunities are realized.

The strongest leaders continuously ask:

What creates value?

What creates waste?

What slows us down?

What improves outcomes?

What should we stop doing?

These questions often uncover significant opportunities for improvement.

The Three Operational Excellence Questions

Before launching any AI initiative, leaders should ask:

Question One

How will this improve productivity?

Question Two

How will this improve quality?

Question Three

How will this improve speed?

If a proposed initiative cannot answer at least one of these questions clearly, its value should be reconsidered.

High-performance organizations maintain relentless focus on outcomes.

Leadership Reflection

Consider your own organization.

- Where does unnecessary friction exist?
- What repetitive work consumes employee time?
- Which processes create delays?
- Where do quality issues occur most frequently?
- What information takes too long to access?
- How could AI improve productivity, quality, or speed?

The answers reveal opportunities for operational transformation.

Key Takeaways

1. Organizations invest in AI to improve performance, not merely adopt technology.
2. Productivity improves when employees spend more time on high-value activities.
3. AI can reduce organizational friction and streamline workflows.
4. Quality improves when errors, risks, and inconsistencies are identified earlier.
5. Speed creates competitive advantage when unnecessary delays are eliminated.
6. AI acts as a capacity multiplier for individuals and teams.
7. Operational excellence remains a leadership responsibility.
8. Success should be measured through outcomes, not technology adoption alone.

The organizations that thrive in the AI era will create measurable improvements in productivity, quality, and speed.

Yet operational excellence alone does not create competitive advantage.

Organizations must also make better decisions.

Because in a world overflowing with information, the quality of decisions increasingly determines the quality of results.

The next chapter explores how leaders can use artificial intelligence to improve judgment, strengthen decision-making, and transform information into action.

Next Chapter: Chapter 16 , AI for Better Decision-Making: From Data to Insight to Action. This chapter demonstrates for leaders how AI enhances decision quality without replacing human judgment.

IN THE
AGE
OF
AI

CHAPTER 16

AI for Better Decision-Making

From Data to Insight to Action

Every result an organization achieves can be traced back to a series of decisions.

Some decisions are small.

Others are significant.

Some affect a single employee.

Others influence customers, investors, and entire communities.

Collectively, decisions shape performance.

They determine growth.

They influence culture.

They affect profitability.

They define success.

This is why decision-making sits at the heart of leadership.

The quality of an organization's decisions ultimately determines the quality of its results.

Artificial intelligence creates an extraordinary opportunity in this area.

Not because it replaces decision-makers.

But because it helps them make better decisions.

The future belongs to organizations that can transform information into insight and insight into action faster and more effectively than their competitors.

The Decision-Making Challenge

Leaders today face a paradox.

They have access to more information than any generation before them.

Yet many decisions remain difficult.

Why?

Because information alone does not create clarity.

In fact, too much information can create confusion.

Executives often find themselves surrounded by:

- Reports.
- Dashboards.
- Metrics.
- Forecasts.
- Market research.
- Customer feedback.
- Financial analyses.

The challenge is no longer obtaining information.

The challenge is understanding what matters.

Artificial intelligence helps by filtering, organizing, analyzing, and prioritizing information.

This allows leaders to focus more energy on judgment rather than information gathering.

The Decision-Making Advantage

Historically, organizations competed through scale.

Through resources.

Through capital.

Through talent.

These factors remain important.

However, in the AI era, decision-making quality is becoming a major source of competitive advantage.

Organizations that consistently make better decisions tend to:

- Respond faster.
- Allocate resources more effectively.
- Serve customers more successfully.
- Innovate more consistently.
- Manage risks more intelligently.

Better decisions create better outcomes.

Repeated over time, those outcomes compound.

The cumulative advantage can be enormous.

Moving Beyond Intuition Alone

Many successful leaders trust their instincts.

Experience matters.

Judgment matters.

Intuition matters.

However, intuition alone has limitations.

Human beings are influenced by:

- Assumptions.
- Biases.
- Emotions.
- Habits.
- Incomplete information.

Artificial intelligence helps provide additional perspective.

It can reveal patterns leaders might miss.

It can identify anomalies.

It can highlight risks.

It can challenge assumptions.

This does not eliminate intuition.

It strengthens it.

The best decisions often emerge when human judgment and data-driven insight work together.

The Decision Pyramid

Effective decision-making can be viewed as a pyramid.

At the base is data.

Above that sits information.

Above information sits insight.

Above insight sits judgment.

At the top sits action.

Many organizations spend enormous energy collecting data.

Far fewer excel at moving up the pyramid.

The goal is not collecting more information.

The goal is creating better actions.

Artificial intelligence helps organizations move more efficiently through the lower levels of the pyramid.

Human leaders remain essential at the upper levels.

AI supports insight.

Humans provide judgment.

Humans take responsibility for action.

Faster Decisions Without Sacrificing Quality

One of the greatest opportunities AI creates is decision speed.

Traditionally, leaders often waited days or weeks to gather sufficient information.

Meetings were scheduled.

Reports were generated.

Analyses were conducted.

Recommendations were developed.

Today, much of that work can occur in minutes.

Information becomes available almost instantly.

Scenarios can be evaluated quickly.

Alternatives can be explored rapidly.

This acceleration allows leaders to spend more time discussing implications rather than collecting facts.

The result is faster decisions without necessarily reducing quality.

In many cases, quality improves because leaders have access to better information.

Predictive Thinking

Another advantage of AI is its ability to support predictive analysis.

Historically, organizations spent significant energy understanding the past.

What happened?

Why did it happen?

What were the results?

These questions remain important.

However, leaders must also think about the future.

What is likely to happen next?

What risks are emerging?

What opportunities are developing?

What outcomes appear most probable?

Artificial intelligence helps organizations move from reactive thinking to proactive thinking.

This creates a significant strategic advantage.

Organizations become better prepared.

They anticipate rather than merely respond.

Better Resource Allocation

Every organization operates with constraints.

Time is limited.

Money is limited.

Talent is limited.

Energy is limited.

One of the most important leadership responsibilities is determining where resources should be invested.

Artificial intelligence can support these decisions by helping leaders identify:

- High-priority opportunities.
- Underperforming activities.
- Emerging risks.
- Resource bottlenecks.
- Areas with the greatest potential return.

This allows organizations to focus attention where it creates the most value.

The result is greater effectiveness.

Not because resources increase.

Because resources are deployed more intelligently.

Decision-Making at Every Level

Many discussions about AI focus on executive decision-making.

The reality is that decisions occur throughout the organization.

Customer service representatives make decisions.

Project managers make decisions.

Sales professionals make decisions.

Supervisors make decisions.

Frontline employees make decisions.

Artificial intelligence can improve decision quality at every level.

Employees gain access to better information.

Recommendations become available.

Relevant knowledge can be retrieved instantly.

The organization becomes smarter collectively.

Decision quality improves throughout the system.

Avoiding Decision Dependency

While AI creates significant benefits, leaders must avoid a potential trap.

Decision dependency.

Decision dependency occurs when people begin relying on technology without applying independent judgment.

Recommendations are accepted automatically.

Outputs go unquestioned.

Critical thinking declines.

This creates risk.

Artificial intelligence should enhance thinking.

Not replace thinking.

Leaders should encourage employees to ask:

Does this recommendation make sense?

What assumptions are being made?

What information may be missing?

What unintended consequences could occur?

Healthy skepticism strengthens decision quality.

The Human Judgment Advantage

The most important decisions often involve factors that cannot be measured easily.

Values.

Relationships.

Culture.

Trust.

Ethics.

Reputation.

Long-term consequences.

Artificial intelligence can provide information.

Human leaders must evaluate these broader considerations.

A recommendation that appears optimal on paper may create unintended consequences in practice.

This is where human judgment remains indispensable.

The organizations that thrive will combine analytical capability with wisdom.

Building a Decision-Making Culture

Technology alone does not improve decisions.

Culture matters.

Organizations should encourage:

- Curiosity.
- Evidence-based thinking.
- Open discussion.
- Constructive challenge.
- Accountability.

People should feel comfortable questioning assumptions.

Sharing perspectives.

Presenting alternative viewpoints.

The strongest decisions often emerge through thoughtful dialogue.

AI can enrich that dialogue.

It cannot replace it.

The Five Questions Great Decision-Makers Ask

High-performance leaders consistently ask:

What problem are we trying to solve?

Clarity precedes quality.

What does the evidence suggest?

Information matters.

What assumptions are we making?

Assumptions influence outcomes.

What are the potential consequences?

Short-term and long-term.

What action creates the greatest value?

The goal is results.

These questions become even more powerful when supported by AI-generated insights.

Leadership Reflection

Ask yourself:

- How quickly does our organization make decisions?
- Do we spend too much time gathering information?
- Are we using evidence effectively?
- Do employees have access to the insights they need?
- Are we encouraging critical thinking?

- How can AI improve decision quality without reducing human judgment?

The answers reveal opportunities for improvement.

Key Takeaways

1. Organizational performance is largely determined by decision quality.
2. AI helps leaders transform information into insight more effectively.
3. Better decisions create sustainable competitive advantage.
4. Human judgment remains essential.
5. AI can improve both decision speed and decision quality.
6. Predictive thinking helps organizations become proactive rather than reactive.
7. Decision-making should improve at every level of the organization.
8. Technology should strengthen critical thinking, not replace it.

The organizations that excel in the AI era will make better decisions faster than their competitors.

They will allocate resources more effectively.

They will anticipate change more successfully.

They will act with greater confidence.

But making good decisions internally is only part of the equation.

Organizations ultimately succeed because they create value for customers.

And in a world where customer expectations continue to rise, artificial intelligence is creating new opportunities to deliver extraordinary experiences.

The next chapter explores how AI is transforming customer experience and helping organizations build deeper, stronger, and more valuable customer relationships.

Next Chapter: Chapter 17 , AI and Customer Experience: Creating More Value for Customers. This chapter focuses on how AI enhances service, personalization, responsiveness, and customer loyalty while preserving the human touch that builds trust.

CHAPTER 17

AI and Customer Experience

Creating More Value for Customers

At the end of the day, every organization exists for one reason:

To create value.

For businesses, value is created for customers.

For nonprofits, value is created for communities and stakeholders.

For government agencies, value is created for citizens.

For educational institutions, value is created for students.

The specific mission may differ, but the principle remains the same.

Organizations survive and thrive when they create meaningful value for the people they serve.

Artificial intelligence is creating new opportunities to do exactly that.

While much of the public discussion around AI focuses on automation and efficiency, some of the greatest opportunities involve improving the customer experience.

Helping customers faster.

Serving customers better.

Understanding customers more deeply.

Creating more personalized experiences.

Delivering greater value.

The organizations that succeed in the coming decade will recognize a simple truth:

AI should not be implemented merely to reduce costs. It should be implemented to increase value.

The Customer Expectation Revolution

Customer expectations have never been static.

They evolve continuously.

Every time an organization improves the customer experience, it raises expectations for everyone else.

Consider what customers now expect:

Fast responses.

Easy access to information.

Personalized recommendations.

Convenient interactions.

Seamless service.

Transparent communication.

These expectations did not exist at the same level twenty years ago.

Today they are commonplace.

Artificial intelligence is accelerating this trend.

Customers increasingly expect organizations to know them, understand them, and respond to their needs quickly and effectively.

Organizations that fail to adapt may find themselves losing customers to competitors that provide more responsive experiences.

The Real Purpose of Customer Experience

Many organizations mistakenly believe customer experience is primarily about being nice.

Being courteous matters.

Being helpful matters.

But customer experience goes deeper.

Customer experience is about making it easier for customers to achieve their desired outcome.

Customers do not buy products.

They buy outcomes.

They buy solutions.

They buy results.

They buy experiences.

The organizations that understand this principle design customer experiences around helping people succeed.

Artificial intelligence becomes valuable when it helps organizations accomplish that objective.

Personalization at Scale

Historically, personalization was difficult.

Small businesses often knew their customers personally.

Large organizations struggled.

As organizations grew, personalization became harder to maintain.

Artificial intelligence changes that dynamic.

Organizations can now analyze customer preferences, behaviors, interactions, and needs at scale.

This creates opportunities to provide:

- Personalized recommendations.
- Customized communication.
- Relevant content.
- Tailored experiences.
- Proactive support.

Customers increasingly appreciate organizations that understand their needs without requiring them to repeat information repeatedly.

Personalization strengthens relationships.

Relationships strengthen loyalty.

Faster Service, Better Service

One of the most visible applications of AI involves customer service.

Customers often become frustrated by:

Long wait times.

Repetitive questions.

Delayed responses.

Complex processes.

Difficulty finding information.

Artificial intelligence can help address many of these challenges.

Customers can receive answers more quickly.

Representatives can access information more easily.

Issues can be routed more effectively.

Knowledge can be retrieved instantly.

The result is often faster service.

However, leaders should remember an important distinction.

Faster service does not automatically mean better service.

The goal is not speed alone.

The goal is creating positive outcomes for customers.

Speed matters because customers value their time.

But quality remains essential.

Empowering Employees to Serve Better

One of the greatest customer experience opportunities is often overlooked.

Artificial intelligence can help employees serve customers more effectively.

Imagine a customer service representative who immediately has access to:

Customer history.

Product information.

Relevant policies.

Suggested solutions.

Previous interactions.

The representative becomes more capable.

More confident.

More effective.

The customer receives a better experience.

The technology supports the employee.

The employee serves the customer.

Everyone benefits.

The best customer experiences often occur when AI enhances human performance rather than replacing human interaction.

Anticipating Customer Needs

Exceptional organizations do more than react.

They anticipate.

They identify needs before customers express them.

They solve problems before they become frustrations.

They create value proactively.

Artificial intelligence helps organizations identify patterns that may indicate:

Emerging concerns.

Future needs.

Service opportunities.

Potential risks.

Changing preferences.

This allows organizations to move from reactive service to proactive service.

Customers notice the difference.

And they remember it.

The Human Touch Still Matters

As AI becomes more prevalent, some organizations may become tempted to automate every interaction.

This would be a mistake.

Technology excels at efficiency.

Humans excel at connection.

Customers often want human interaction when:

Problems are complex.

Emotions are involved.

Trust is important.

Situations are unique.

Relationships matter.

The strongest organizations understand when technology should lead and when people should lead.

The goal is balance.

Technology handles routine interactions.

People handle moments that require empathy, judgment, and relationship-building.

This combination often produces the best results.

Building Loyalty Through Value

Customer loyalty is earned.

It cannot be purchased.

Customers remain loyal when organizations consistently create value.

Artificial intelligence can support loyalty by helping organizations:

- Understand customers better.
- Communicate more effectively.
- Resolve issues faster.
- Deliver more relevant experiences.
- Improve consistency.

However, loyalty ultimately depends on trust.

Organizations that misuse technology, violate privacy, or create frustrating experiences damage trust.

Organizations that use technology responsibly strengthen it.

Trust remains the foundation of loyalty.

Measuring Customer Experience

Many organizations measure activity.

Fewer measure outcomes.

When evaluating customer experience initiatives, leaders should focus on metrics such as:

- Customer satisfaction.
- Customer retention.
- Customer loyalty.
- Resolution speed.
- Response quality.
- Repeat business.
- Referral rates.

Technology adoption is not the goal.

Customer value is the goal.

Every initiative should ultimately answer one question:

How does this improve the customer experience?

If the answer is unclear, the initiative should be reconsidered.

Creating Customer-Centered Organizations

Technology alone does not create great customer experiences.

Culture does.

Organizations must build cultures that genuinely care about customer success.

Employees should understand:

Who the customer is.

What the customer values.

What challenges the customer faces.

How the organization creates value.

Artificial intelligence becomes most effective when it operates within a customer-centered culture.

Technology amplifies the organization's intentions.

If the organization genuinely seeks to help customers, AI can significantly enhance those efforts.

The Competitive Advantage of Customer Understanding

Many organizations focus heavily on products.

Others focus on operations.

The strongest organizations focus on understanding customers.

Why?

Because customer understanding drives everything else.

It informs innovation.

It improves service.

It strengthens relationships.

It creates loyalty.

Artificial intelligence provides organizations with unprecedented opportunities to understand customers at a deeper level.

The organizations that leverage this capability effectively will create significant competitive advantages.

Leadership Reflection

Ask yourself:

- How well do we understand our customers?
- Are we using AI to create value or merely reduce costs?
- Where are customers experiencing friction?
- How can we improve responsiveness?
- Are employees empowered to serve customers effectively?
- Are we preserving the human touch where it matters most?

The answers reveal opportunities for growth.

Key Takeaways

1. Organizations exist to create value for those they serve.
2. AI should be used to increase customer value, not simply reduce costs.
3. Customer expectations continue to rise as technology advances.
4. Personalization at scale is becoming increasingly possible.
5. Faster service is valuable when it improves outcomes.
6. AI can empower employees to deliver better customer experiences.
7. Human connection remains essential.
8. Trust and value are the foundations of customer loyalty.

The organizations that win in the AI era will not merely operate more efficiently.

They will create better experiences.

They will understand customers more deeply.

They will build stronger relationships.

And they will consistently deliver greater value.

Yet serving customers well today is only part of the challenge.

Organizations must also prepare for tomorrow.

They must innovate.

They must experiment.

They must discover new ways to create value.

The next chapter explores how artificial intelligence can accelerate innovation and help organizations unlock opportunities that were previously beyond reach.

Next Chapter: Chapter 18 , Innovation at Scale: Using AI to Accelerate Growth and Opportunity.

This chapter illustrates for leaders how AI can become a catalyst for creativity, experimentation, product development, and organizational growth.

IN THE
AGE
OF
AI

CHAPTER 18

Innovation at Scale

Using AI to Accelerate Growth and Opportunity

Every organization faces the same challenge:

How do we continue creating value tomorrow while succeeding today?

Operational excellence helps organizations perform efficiently.

Customer experience helps organizations create loyalty.

Leadership helps organizations navigate change.

Innovation helps organizations remain relevant.

Without innovation, organizations eventually stagnate.

Products become outdated.

Services become commoditized.

Competitors gain ground.

Opportunities are missed.

History is filled with examples of successful organizations that failed because they became too attached to the past.

They continued doing what had worked yesterday while the world moved forward.

Artificial intelligence is creating an opportunity to avoid that fate.

Not because AI replaces innovation.

But because it accelerates it.

The organizations that learn how to combine human creativity with artificial intelligence will discover opportunities that were previously difficult, or impossible, to uncover.

Innovation Is Not Optional

Many leaders treat innovation as a luxury.

Something to focus on when time permits.

When budgets allow.

When operations are running smoothly.

This mindset creates risk.

Innovation is no longer optional.

In rapidly changing markets, innovation is essential to survival.

Customer expectations evolve.

Technology advances.

Competitors emerge.

New business models appear.

Organizations that fail to innovate eventually find themselves reacting rather than leading.

The strongest organizations build innovation into their culture, strategy, and daily operations.

They make innovation a discipline.

Not an event.

The Innovation Myth

One of the biggest misconceptions about innovation is that it requires rare genius.

Many people imagine innovation as dramatic breakthroughs created by extraordinary individuals.

While breakthroughs occur, most innovation is far less dramatic.

It often involves:

- Improving a process.
- Solving a customer problem.
- Eliminating friction.
- Combining existing ideas in new ways.
- Creating a better experience.
- Discovering a new opportunity.

Innovation happens when people ask better questions.

Notice unmet needs.

Challenge assumptions.

Explore possibilities.

Artificial intelligence can support each of these activities.

Expanding Human Creativity

Some people worry that AI will reduce creativity.

In practice, many organizations are discovering the opposite.

Artificial intelligence can serve as a creative partner.

It can generate ideas.

Suggest alternatives.

Challenge assumptions.

Explore possibilities.

Provide different perspectives.

This does not replace human creativity.

It expands it.

The human imagination remains essential.

People define the problem.

People evaluate ideas.

People exercise judgment.

People determine which opportunities are worth pursuing.

AI simply helps accelerate the process.

The result is often more creativity, not less.

Accelerating Experimentation

Innovation requires experimentation.

Organizations learn by trying.

Testing.

Exploring.

Improving.

Historically, experimentation could be expensive and time-consuming.

Artificial intelligence helps reduce those barriers.

Ideas can be explored faster.

Scenarios can be evaluated more quickly.

Prototypes can be developed more efficiently.

Customer feedback can be analyzed rapidly.

The speed of learning increases.

This is important because innovation is fundamentally a learning process.

Organizations that learn faster often innovate faster.

Organizations that innovate faster often create greater value.

The Innovation Pipeline

High-performance organizations treat innovation like a pipeline.

Ideas enter.

Experiments occur.

Insights emerge.

Opportunities advance.

Weak ideas are discarded.

Strong ideas are refined.

Over time, innovation becomes systematic rather than accidental.

Artificial intelligence can support every stage of the pipeline.

It can help identify opportunities.

Analyze trends.

Generate alternatives.

Evaluate possibilities.

Monitor outcomes.

The organization becomes more capable of discovering and developing opportunities consistently.

Finding Hidden Opportunities

One of the greatest strengths of AI is its ability to identify patterns.

Patterns often reveal opportunities.

Customer complaints may reveal unmet needs.

Market trends may reveal emerging demand.

Operational inefficiencies may reveal improvement opportunities.

Behavioral data may reveal new business possibilities.

Historically, many of these patterns remained hidden because organizations lacked the ability to analyze large amounts of information effectively.

Today, those opportunities can become visible.

The challenge shifts from finding information to acting on insight.

Innovation Beyond Products

When leaders hear the word innovation, they often think about products.

Products matter.

But innovation can occur throughout the organization.

Organizations can innovate:

- Processes.
- Services.
- Customer experiences.
- Business models.
- Partnerships.
- Workforce development.
- Communication systems.
- Organizational structures.

Some of the most valuable innovations are operational rather than technological.

The goal is creating value.

Innovation simply provides another path to achieve it.

Creating a Culture of Innovation

Technology alone does not create innovation.

Culture does.

Employees must feel encouraged to:

Ask questions.

Challenge assumptions.

Suggest improvements.

Explore possibilities.

Experiment responsibly.

Learn from failure.

Organizations that punish mistakes often suppress innovation.

Organizations that encourage learning tend to accelerate it.

Leaders play a critical role.

They determine whether curiosity is rewarded or discouraged.

Whether experimentation is supported or avoided.

Whether innovation becomes part of the culture or remains isolated.

The Growth Connection

Innovation and growth are closely linked.

Organizations grow when they create new value.

New value often emerges through innovation.

Artificial intelligence helps organizations identify opportunities for growth by improving their ability to:

Understand customers.

Analyze markets.

Evaluate opportunities.

Develop solutions.

Scale successful initiatives.

Growth becomes more intentional.

Less dependent on luck.

More dependent on capability.

Innovation and Risk

Some leaders hesitate to pursue innovation because they fear risk.

The concern is understandable.

Every innovation involves uncertainty.

However, refusing to innovate creates risk as well.

The risk of irrelevance.

The risk of stagnation.

The risk of being overtaken by competitors.

Artificial intelligence can help reduce certain innovation risks by improving analysis and accelerating learning.

Organizations gain information more quickly.

Experiments become less expensive.

Feedback arrives faster.

Risk does not disappear.

But it becomes more manageable.

The Innovation Leadership Mindset

Leaders who excel at innovation share several characteristics.

They remain curious.

They ask questions.

They challenge assumptions.

They encourage exploration.

They tolerate intelligent risk-taking.

They focus on learning.

Most importantly, they believe improvement is always possible.

This mindset influences the entire organization.

Employees follow the example set by leadership.

If leaders remain curious, innovation grows.

If leaders become complacent, innovation slows.

The Future Belongs to Builders

The AI era creates enormous opportunities.

Some organizations will use AI primarily to improve efficiency.

Others will use it to build.

Build better experiences.

Build new products.

Build stronger organizations.

Build new markets.

Build new possibilities.

The greatest rewards often go to builders.

Organizations willing to imagine, experiment, and create.

Technology expands what is possible.

Leadership determines what becomes reality.

Leadership Reflection

Ask yourself:

- How effectively does our organization innovate today?
- Are employees encouraged to experiment?
- Do we treat innovation as a discipline or an occasional activity?
- How can AI accelerate learning and experimentation?
- Where might hidden opportunities exist?
- Are we creating the future or reacting to it?

The answers reveal the organization's innovation readiness.

Key Takeaways

1. Innovation is essential for long-term relevance and growth.
2. AI accelerates innovation by expanding creativity and reducing barriers to experimentation.
3. Most innovation involves solving problems and creating value rather than dramatic breakthroughs.
4. Organizations should build systematic innovation pipelines.
5. Hidden opportunities often emerge through pattern recognition and analysis.
6. Innovation can occur across products, services, processes, and business models.
7. Culture plays a critical role in innovation success.
8. The organizations that thrive will use AI not only to improve efficiency but also to create new opportunities.

The organizations that succeed in the AI era will not simply operate better.

They will imagine better.

They will learn faster.

They will innovate continuously.

And they will create opportunities where others see disruption.

But innovation alone does not guarantee success.

Organizations must also translate innovation into growth.

They must create new revenue streams, expand markets, strengthen competitiveness, and generate sustainable value.

The next chapter explores how leaders can use artificial intelligence not only to improve performance, but also to unlock entirely new avenues for growth and expansion.

Next Chapter: Chapter 19 , The AI-Enhanced Executive: How Leaders Can Use AI as a Strategic Thought Partner.

This chapter specifically outlines practical ways for CEOs, executives, consultants, and business owners to personally use AI every day. This will help to galvanize the powerful concepts presented within the pages of this book as immediately useful and highly referable.

IN THE
AGE
OF
AI

CHAPTER 19

The AI-Enhanced Executive

How Leaders Can Use AI as a Strategic Thought Partner

One of the most significant opportunities created by artificial intelligence has nothing to do with automation.

It has nothing to do with customer service.

It has nothing to do with operational efficiency.

It has everything to do with leadership.

For the first time in history, executives have access to a tool that can function as a strategic thought partner twenty-four hours a day, seven days a week.

Not a replacement for judgment.

Not a replacement for experience.

Not a replacement for wisdom.

A thought partner.

A sounding board.

A research assistant.

A brainstorming partner.

A planning resource.

A communication coach.

A decision-support tool.

The executives who learn to use AI effectively may gain one of the most significant productivity and leadership advantages of the coming decade.

The reason is simple.

Leadership is fundamentally a thinking profession.

And AI can help leaders think more effectively.

The Executive Time Challenge

Most executives face the same problem.

Too much information.

Too many meetings.

Too many decisions.

Too many competing priorities.

Too little time.

The challenge is rarely effort.

Most leaders already work hard.

The challenge is focus.

How do leaders spend more time thinking strategically and less time managing information?

Artificial intelligence can help.

By reducing time spent on routine cognitive tasks, leaders gain more capacity for high-value thinking.

The result is not simply greater efficiency.

It is greater effectiveness.

AI as a Thinking Partner

Many executives initially use AI as a search tool.

A writing tool.

A research tool.

These applications are valuable.

The greater opportunity lies in using AI as a thinking partner.

For example, leaders can ask AI to:

Challenge assumptions.

Identify risks.

Generate alternatives.

Explore scenarios.

Evaluate options.

Prepare questions.

Analyze strategies.

Surface blind spots.

This creates a powerful dynamic.

The executive brings experience, context, judgment, and organizational knowledge.

AI brings speed, information processing, and perspective.

Together, they create stronger thinking.

Strategic Planning

Strategic planning often requires leaders to consider multiple possibilities.

Future trends.

Competitive threats.

Market shifts.

Customer expectations.

Economic conditions.

Artificial intelligence can accelerate this process.

Executives can use AI to:

- Explore strategic options.
- Conduct scenario planning.
- Analyze industry trends.
- Identify emerging opportunities.
- Evaluate potential risks.

Rather than staring at a blank page, leaders begin with a set of possibilities.

This often improves both the quality and speed of strategic planning.

Decision Preparation

One of the most valuable executive uses of AI is decision preparation.

Consider the number of decisions leaders make every day.

Hiring decisions.

Investment decisions.

Partnership decisions.

Operational decisions.

Strategic decisions.

AI can help organize information, summarize alternatives, identify considerations, and highlight potential consequences.

This allows executives to focus on judgment rather than information gathering.

The leader remains accountable.

The decision remains human.

The preparation becomes more effective.

Executive Communication

Communication is one of the most important leadership responsibilities.

Leaders communicate through:

Emails.

Presentations.

Speeches.

Meetings.

Reports.

Town halls.

Customer messages.

Board updates.

Artificial intelligence can help executives:

Draft messages.

Improve clarity.

Refine language.

Adapt communication to different audiences.

Generate presentation outlines.

Prepare talking points.

Develop leadership messages.

This does not replace the leader's voice.

It strengthens it.

The executive still provides vision, authenticity, and purpose.

AI helps improve delivery.

Meeting Effectiveness

Many executives spend enormous portions of their day in meetings.

Some meetings create value.

Others consume time.

Artificial intelligence can improve meeting effectiveness by helping leaders:

Prepare agendas.

Identify key questions.

Summarize discussions.

Capture action items.

Track commitments.

Review outcomes.

The result is more productive conversations and stronger follow-through.

The organization benefits because decisions move faster and accountability improves.

Problem Solving

Every organization encounters problems.

Operational issues.

Customer concerns.

Personnel challenges.

Market disruptions.

Strategic obstacles.

Artificial intelligence can support structured problem solving by helping leaders:

Analyze root causes.

Generate alternatives.

Explore solutions.

Evaluate trade-offs.

Identify potential risks.

The technology does not solve the problem.

It helps leaders think more comprehensively about the problem.

That distinction matters.

The Executive Learning Advantage

One of the greatest benefits of AI is accelerated learning.

Executives can quickly gain familiarity with:

New industries.

Emerging technologies.

Market developments.

Leadership concepts.

Regulatory changes.

Business models.

The speed of learning increases dramatically.

This creates a competitive advantage.

Leaders become better informed.

Better informed leaders often make better decisions.

The ability to learn quickly may become one of the most important executive capabilities of the future.

AI and Board Leadership

Boards increasingly ask questions about artificial intelligence.

Executives must be prepared.

They need to discuss:

Opportunities.

Risks.

Investments.

Governance.

Workforce implications.

Competitive strategy.

AI can help leaders prepare for these conversations by organizing information, identifying key considerations, and generating discussion frameworks.

The result is more productive board engagement and stronger governance.

Avoiding the Executive Trap

Despite its benefits, AI introduces a potential danger.

Overreliance.

Some executives may become tempted to outsource too much thinking.

This is a mistake.

Artificial intelligence can support thinking.

It cannot replace leadership.

Leadership requires:

Judgment.

Values.

Experience.

Relationships.

Context.

Wisdom.

The executive who delegates thinking entirely to technology weakens the very capabilities that make leadership valuable.

The strongest leaders use AI to expand their thinking, not replace it.

The Daily Executive AI Framework

Executives can create tremendous value by using AI in five ways every day.

Think

Use AI to explore ideas, assumptions, and possibilities.

Learn

Use AI to accelerate knowledge acquisition.

Plan

Use AI to organize priorities and strategies.

Communicate

Use AI to improve clarity and effectiveness.

Reflect

Use AI to challenge thinking and uncover blind spots.

This framework helps leaders integrate AI into daily practice without becoming dependent on it.

The Multiplication Effect

The true value of AI is not automation.

It is multiplication.

Multiplying:

Thinking.

Learning.

Creativity.

Productivity.

Preparation.

Communication.

Decision-making.

Leadership capacity.

A leader who uses AI effectively may accomplish significantly more than a leader who does not.

Not because the leader works harder.

Because the leader works smarter.

The Executive of the Future

The executive of the future will likely possess three characteristics.

First, they will remain deeply human.

They will build trust.

Develop people.

Create culture.

Provide vision.

Second, they will be highly adaptable.

They will learn continuously.

Challenge assumptions.

Embrace change.

Third, they will leverage intelligent tools effectively.

They will combine human judgment with machine intelligence.

This combination creates extraordinary leadership capability.

Leadership Reflection

Ask yourself:

- How am I personally using AI today?
- Am I using AI strategically or only tactically?
- Could AI help me think more effectively?
- Could it improve communication?
- Could it accelerate learning?
- Could it help me identify blind spots?

The answers may reveal opportunities to become a more effective leader.

Key Takeaways

1. AI can function as a powerful executive thought partner.
2. Leadership remains fundamentally a human responsibility.
3. AI helps leaders think, learn, plan, communicate, and solve problems more effectively.

4. Strategic planning and decision preparation can be significantly enhanced through AI.
5. Communication effectiveness improves when AI supports message development.
6. Executives should use AI to expand thinking, not replace thinking.
7. Continuous learning becomes a critical leadership advantage.
8. The future belongs to leaders who successfully combine human wisdom with intelligent tools.

Artificial intelligence is changing how leaders think, communicate, and make decisions.

But leadership effectiveness ultimately serves a larger purpose.

Organizations exist to create value.

And when leaders combine operational excellence, innovation, customer focus, and intelligent decision-making, they unlock something every organization seeks:

Growth.

The next chapter explores how artificial intelligence can help organizations create new opportunities, expand markets, generate revenue, and build sustainable growth in the years ahead.

Next Chapter: Chapter 20 , AI and Growth: Creating New Revenue, Markets, and Opportunities. This chapter transitions into future-focused discussions, and concludes with *the AI Transformation Roadmap.*

PART V
THE FUTURE

IN THE
AGE OF AI

CHAPTER 20

AI and Growth

Creating New Revenue, Markets, and Opportunities

Every organization eventually reaches a point where improving efficiency is no longer enough.

Processes have been optimized.

Costs have been reduced.

Productivity has improved.

Operations have become more effective.

The next question becomes:

How do we grow?

Growth is the lifeblood of most organizations.

Growth creates opportunity.

Growth attracts talent.

Growth funds innovation.

Growth increases impact.

Growth creates resources that allow organizations to pursue their mission more effectively.

Artificial intelligence is creating unprecedented opportunities for growth.

Not simply by helping organizations do existing things better.

But by helping them discover entirely new ways to create value.

The organizations that thrive in the coming decade will use AI not only as an efficiency tool but as a growth engine.

The Growth Mindset

Before discussing technology, it is important to discuss mindset.

Organizations generally approach AI from one of two perspectives.

The first group asks:

How can we reduce costs?

The second group asks:

How can we create more value?

Both questions matter.

However, growth-oriented organizations focus primarily on the second.

Cost reduction creates short-term benefits.

Value creation creates long-term opportunities.

The leaders who generate extraordinary growth often spend less time asking:

"What can we eliminate?"

And more time asking:

"What can we build?"

Artificial intelligence expands the possibilities available to builders.

Growth Through Customer Understanding

The first source of growth is customer understanding.

Organizations often believe they know their customers.

Sometimes they do.

Often, they know only part of the story.

Artificial intelligence helps organizations identify:

- Customer preferences.
- Behavioral patterns.
- Emerging needs.
- Buying trends.
- Service opportunities.
- Retention risks.

These insights create opportunities for growth.

When organizations understand customers more deeply, they can:

Serve them better.

Solve more problems.

Create additional value.

Develop stronger relationships.

Growth frequently begins with understanding.

Discovering New Markets

Many organizations focus almost exclusively on their existing markets.

Artificial intelligence can help leaders identify opportunities beyond their traditional boundaries.

New geographic markets.

New customer segments.

New industries.

New applications.

New partnerships.

Historically, discovering these opportunities often required extensive research and significant resources.

Today, organizations can analyze trends and possibilities far more efficiently.

The result is broader visibility and more informed growth decisions.

Growth opportunities that once remained hidden become visible.

Product and Service Expansion

Artificial intelligence can help organizations identify unmet needs.

Those unmet needs often become the foundation for new products and services.

Customers frequently communicate opportunities without realizing it.

Questions they ask.

Problems they face.

Frustrations they experience.

Workarounds they create.

All of these provide clues.

Organizations that listen carefully often discover opportunities competitors miss.

Artificial intelligence helps organizations analyze large amounts of customer feedback and identify recurring themes.

These insights become innovation opportunities.

Innovation becomes growth.

Personalization as a Growth Driver

Customers increasingly expect personalized experiences.

They want relevant recommendations.

Relevant communication.

Relevant solutions.

Artificial intelligence allows organizations to deliver personalization at a scale previously impossible.

This creates multiple growth opportunities.

Increased customer engagement.

Higher retention.

Improved customer satisfaction.

Additional revenue opportunities.

Stronger loyalty.

Organizations that personalize effectively often create deeper customer relationships and stronger long-term growth.

Accelerating Sales Performance

Sales organizations are already experiencing significant benefits from AI.

Sales professionals can:

Identify opportunities faster.

Prepare more effectively.

Prioritize prospects.

Personalize outreach.

Analyze customer needs.

Improve follow-up.

Forecast results.

The result is not simply increased activity.

It is increased effectiveness.

The best sales organizations use AI to improve the quality of customer interactions rather than merely increasing the quantity.

Growth follows value creation.

Expanding Organizational Capacity

Growth often creates challenges.

More customers.

More transactions.

More complexity.

More operational demands.

Historically, organizations often struggled to scale efficiently.

Artificial intelligence creates new possibilities.

Organizations can expand capacity without increasing resources proportionally.

Customer support becomes more scalable.

Operations become more efficient.

Information becomes more accessible.

Decision-making becomes faster.

This allows organizations to pursue growth more confidently.

The Opportunity Creation Mindset

One of the most powerful applications of AI is opportunity creation.

Many leaders spend most of their time solving problems.

Problem-solving is important.

However, growth often emerges from opportunity-seeking.

Organizations should regularly ask:

What opportunities are emerging?

What customer needs remain unmet?

What partnerships could be developed?

What markets could be served?

What services could be created?

Artificial intelligence helps organizations explore these questions more effectively.

The result is a broader view of what is possible.

Business Model Innovation

Some of the most significant growth opportunities arise through business model innovation.

Organizations often assume growth requires doing more of what they already do.

Sometimes it does.

Other times, growth requires doing something entirely different.

Artificial intelligence enables organizations to reconsider:

How value is delivered.

How services are packaged.

How customers are served.

How revenue is generated.

How relationships are maintained.

Organizations willing to rethink their business models often discover opportunities competitors overlook.

Scaling Knowledge

Knowledge has traditionally been difficult to scale.

The expertise of a top salesperson.

The wisdom of an experienced manager.

The insights of a seasoned executive.

These capabilities often remained concentrated in individuals.

Artificial intelligence creates opportunities to distribute knowledge more broadly.

Best practices can be captured.

Expertise can be shared.

Insights can be accessed more easily.

The organization becomes smarter collectively.

This creates a significant growth advantage.

Growth Through Speed

Growth opportunities often have a limited window.

Organizations that identify opportunities early gain advantages.

Organizations that move slowly frequently miss them.

Artificial intelligence helps accelerate:

Research.

Analysis.

Decision-making.

Execution.

The result is greater organizational agility.

Organizations become more capable of responding to opportunities as they emerge.

Speed becomes a growth advantage.

Sustainable Growth

Not all growth is good growth.

Organizations sometimes pursue growth that creates complexity without value.

Revenue without profitability.

Expansion without capability.

Scale without quality.

High-performance organizations pursue sustainable growth.

Growth aligned with purpose.

Growth aligned with culture.

Growth aligned with customer value.

Artificial intelligence should support sustainable growth rather than growth at any cost.

The strongest organizations remain disciplined.

They grow intentionally.

The Growth Leadership Challenge

Growth requires leadership.

Technology can reveal opportunities.

People must pursue them.

Leaders must:

Set direction.

Allocate resources.

Build capability.

Create alignment.

Manage risk.

Inspire execution.

Artificial intelligence expands what leaders can see.

Leadership determines what organizations do with what they see.

Leadership Reflection

Ask yourself:

- Where are our greatest growth opportunities?
- Are we using AI to create value or merely reduce costs?
- How well do we understand our customers?
- What new markets could we explore?
- What products or services might customers need next?
- Are we positioned to scale successfully?

The answers reveal the organization's growth readiness.

Key Takeaways

1. Growth is ultimately driven by value creation.
2. AI creates opportunities to identify new markets, customers, and revenue streams.
3. Customer understanding is often the starting point for growth.
4. Personalization can strengthen engagement, loyalty, and revenue.
5. AI improves sales effectiveness and organizational scalability.
6. Opportunity creation is as important as problem-solving.

7. Business model innovation can unlock significant growth.
8. Sustainable growth requires alignment with purpose, culture, and customer value.

The organizations that thrive in the AI era will not simply become more efficient.

They will become more expansive.

They will discover new opportunities.

Serve more customers.

Create more value.

And generate greater impact.

Yet growth alone is not the ultimate objective.

The larger question remains:

What will high-performance organizations look like in the future?

As artificial intelligence continues evolving, how will leadership, work, culture, and organizational design evolve alongside it?

The next chapter explores the future and what leaders should begin preparing for today.

Next Chapter: Chapter 21 , What High-Performance Organizations Will Look Like in 2035. This chapter paints a compelling vision of the future organization.

IN THE
AGE
OF
AI

CHAPTER 21

What High-Performance Organizations Will Look Like in 2035

Preparing for the Next Decade

Predicting the future is risky.

History is filled with experts who confidently forecasted outcomes that never occurred.

Technologies emerged unexpectedly.

Industries evolved differently than anticipated.

Consumer behavior changed.

New opportunities appeared.

Old assumptions disappeared.

The purpose of this chapter is not to predict the future with certainty.

The purpose is to identify trends that are already emerging and explore what they may mean for leaders and organizations over the next decade.

While no one knows exactly what 2035 will look like, one thing seems increasingly clear:

The organizations that thrive will look very different from many organizations today.

Not because technology changes.

Because the nature of work, leadership, learning, and value creation continues to evolve.

The future belongs to organizations that begin preparing now.

The Learning Organization Becomes the Dominant Organization

Throughout this book, one theme has appeared repeatedly:

Learning.

The organizations of the future will not necessarily be the largest.

The richest.

Or even the most technologically advanced.

They will be the fastest learners.

Why?

Because technology evolves continuously.

Customer expectations change rapidly.

Market conditions shift unexpectedly.

Competitive advantages become temporary.

In this environment, learning becomes the ultimate strategic capability.

Organizations that learn faster will adapt faster.

Organizations that adapt faster will outperform competitors.

By 2035, learning may become as important as finance, operations, marketing, and strategy.

The organization itself becomes a learning system.

Human-AI Collaboration Becomes Normal

Today, many organizations are still experimenting with AI.

By 2035, human-AI collaboration will likely become routine.

Employees will interact with intelligent systems throughout the day.

Not as occasional tools.

As integrated work partners.

AI may help employees:

Prepare for meetings.

Analyze information.

Develop ideas.

Create content.

Solve problems.

Manage projects.

Support customers.

The technology will become less visible and more embedded.

Much like electricity today.

People no longer discuss electricity constantly because it is integrated into daily life.

Artificial intelligence may follow a similar path.

The focus will shift from the technology itself to the value it creates.

Leadership Evolves Further

Leadership will remain essential.

Perhaps more essential than ever.

However, leadership responsibilities will continue evolving.

Future leaders will spend less time managing information and more time creating alignment.

Less time supervising activities and more time developing capabilities.

Less time controlling work and more time enabling performance.

The future leader becomes:

A coach.

A teacher.

A strategist.

A culture builder.

A talent developer.

An architect of organizational success.

Leadership becomes increasingly human as technology becomes increasingly capable.

Organizational Structures Become More Flexible

Many traditional organizational structures were designed for stability.

The future demands adaptability.

Organizations are likely to become:

More agile.

More collaborative.

More cross-functional.

More project-based.

More networked.

Departments will continue to exist.

However, collaboration across boundaries will become increasingly important.

Work will flow more fluidly.

Teams will form and reform based on priorities.

The ability to adapt quickly becomes a competitive advantage.

Skills Have Shorter Lifespans

Historically, employees could rely on the same skills for many years.

That reality is changing.

By 2035, skills may have significantly shorter useful lives.

Employees will need to:

Learn continuously.

Develop new capabilities.

Adapt repeatedly.

Reinvent themselves periodically.

This is not a threat.

It is an opportunity.

Organizations that support continuous learning will attract and retain stronger talent.

Employees who embrace lifelong learning will create more opportunities for themselves.

Learning becomes the new job security.

Data Becomes Organizational Intelligence

Today, many organizations still struggle with fragmented information.

In the future, information will become increasingly integrated.

Organizations will possess greater visibility into:

Customers.

Operations.

Markets.

Workforce performance.

Emerging opportunities.

Decision-making becomes more informed.

Insights become more accessible.

Learning accelerates.

The organizations that treat data strategically will possess a significant advantage.

Not because they have more information.

Because they use information more effectively.

Customer Expectations Continue Rising

Customer expectations never move backward.

Every improvement creates a new standard.

By 2035, customers will likely expect:

Faster service.

More personalization.

Greater convenience.

More proactive support.

Seamless experiences.

Organizations that cannot deliver these experiences may struggle.

The strongest organizations will continuously improve how they create value for customers.

Technology enables this.

Customer focus drives it.

Trust Becomes More Valuable

As technology becomes more powerful, trust becomes more important.

Customers will increasingly ask:

Can I trust this organization?

Employees will ask:

Can I trust this leadership team?

Communities will ask:

Can I trust how this technology is being used?

Trust will influence:

Customer loyalty.

Employee engagement.

Brand reputation.

Organizational performance.

Organizations that build trust intentionally will create lasting advantages.

Organizations that neglect trust may find recovery difficult.

Purpose Becomes a Differentiator

The future will not be won by technology alone.

Technology becomes increasingly accessible.

Purpose remains unique.

People want meaning.

Employees want purpose.

Customers want organizations they believe in.

Communities want organizations that contribute positively.

The strongest organizations will combine:

Purpose.

People.

Technology.

Performance.

These elements reinforce one another.

Purpose provides direction.

Technology provides capability.

People provide execution.

Performance creates impact.

The Rise of the Adaptive Organization

Perhaps the defining characteristic of future organizations will be adaptability.

The ability to:

Learn.

Adjust.

Experiment.

Improve.

Respond.

Reinvent.

Adaptability becomes more valuable than stability.

Organizations that cling too tightly to old models may struggle.

Organizations that evolve continuously gain advantage.

The adaptive organization becomes the high-performance organization.

The Future Workforce

The workforce of 2035 will likely value:

Growth.

Flexibility.

Purpose.

Learning.

Meaningful work.

Development opportunities.

Employees will increasingly seek organizations that help them grow.

Not simply organizations that provide employment.

This creates both a challenge and an opportunity for leaders.

Organizations that invest in people will attract stronger talent.

Organizations that neglect development may struggle to compete.

The Organizations That Win

When we step back and examine all of these trends, a picture begins to emerge.

The organizations that thrive by 2035 will likely share several characteristics.

They will:

Learn continuously.

Develop people aggressively.

Leverage AI effectively.

Build trust intentionally.

Adapt rapidly.

Innovate consistently.

Serve customers exceptionally.

Lead with purpose.

These characteristics matter more than any specific technology.

Because technologies change.

Principles endure.

Leadership Reflection

Ask yourself:

- Is our organization built for adaptation?
- Are we investing enough in learning?
- Are we developing future-ready leaders?
- Are we building trust intentionally?
- Are we preparing employees for continuous change?
- Would our organization thrive in 2035 if nothing changed from today?

The answers reveal how prepared the organization is for the future.

Key Takeaways

1. The future belongs to learning organizations.
2. Human-AI collaboration will become increasingly normal.
3. Leadership will become more focused on people, culture, and capability development.

4. Organizational structures will become more adaptive and collaborative.
5. Continuous learning will become essential for both individuals and organizations.
6. Trust will become an increasingly important competitive advantage.
7. Purpose will remain a powerful differentiator.
8. Adaptability may become the defining characteristic of high-performance organizations.

The future is not something that happens to organizations.

It is something organizations create.

The leaders who begin preparing today will be positioned to thrive tomorrow.

And while technology will continue evolving, one truth remains constant:

Organizations succeed because of people.

The final chapter explores why the human advantage remains the ultimate competitive advantage, and how leaders can ensure that technology strengthens, rather than diminishes, what makes organizations extraordinary.

Next Chapter: Chapter 22 , The Human Advantage: Why Leadership, Purpose, and Culture Will Still Matter Most. This serves as the capstone chapter before the conclusion and AI Transformation Roadmap.

IN THE
AGE
OF
AI

CHAPTER 22

The Human Advantage

Why Leadership, Purpose, and Culture Will Still Matter Most

As we reach the conclusion of this journey, it is worth returning to the question that has quietly followed us throughout this book:

What will matter most in the age of artificial intelligence?

Technology will continue evolving.

Algorithms will become more capable.

Systems will become more intelligent.

Automation will become more sophisticated.

Organizations will gain access to tools that seem remarkable by today's standards.

Yet despite all of these advancements, one truth remains unchanged:

Organizations are, and always will be, human enterprises.

Technology may amplify performance.

People create meaning.

Technology may improve efficiency.

People create purpose.

Technology may accelerate execution.

People create vision.

Technology may generate information.

People create wisdom.

The future belongs to organizations that understand this distinction.

The Greatest Misunderstanding About AI

One of the biggest misconceptions about artificial intelligence is the belief that technology itself creates competitive advantage.

For a brief period, it might.

Eventually, however, most technologies become widely available.

Competitors gain access.

Costs decrease.

Capabilities spread.

What once created differentiation becomes commonplace.

The same pattern is likely to occur with AI.

The tools will become more accessible.

The playing field will level.

When that happens, leaders will discover something important:

The true competitive advantage was never the technology.

It was the organization's ability to use the technology effectively.

That ability comes from leadership.

Culture.

Trust.

Learning.

Innovation.

Purpose.

People.

These factors remain difficult to replicate.

Leadership Remains the Multiplier

Throughout history, great organizations have rarely emerged because of technology alone.

They emerged because of leadership.

Leaders establish direction.

Leaders create clarity.

Leaders build trust.

Leaders develop people.

Leaders shape culture.

Leaders create environments where extraordinary performance becomes possible.

Artificial intelligence changes many things.

It does not change the need for leadership.

If anything, it increases it.

Periods of rapid change create greater demand for leaders who can:

Navigate uncertainty.

Communicate effectively.

Make sound decisions.

Inspire confidence.

Develop capability.

Build resilient organizations.

Technology may accelerate the pace of change.

Leadership determines how successfully organizations respond.

Culture Is Still the Operating System

Earlier in this book, we described culture as the operating system of the organization.

That description becomes even more important in the AI era.

Consider two organizations with access to identical technology.

One organization possesses:

High trust.

Strong communication.

Continuous learning.

Shared accountability.

Collaborative leadership.

The other possesses:

Low trust.

Poor communication.

Political behavior.

Resistance to change.

Limited accountability.

Which organization will generate better results?

The answer is obvious.

Technology amplifies culture.

It does not replace it.

Healthy cultures become stronger.

Unhealthy cultures become more exposed.

This is why leaders must invest in culture intentionally.

It remains one of the most powerful drivers of long-term performance.

Purpose Creates Endurance

Organizations without purpose often struggle during periods of disruption.

Why?

Because purpose provides stability when circumstances change.

Markets evolve.

Technology advances.

Strategies shift.

Purpose endures.

Purpose answers questions such as:

Why do we exist?

Who do we serve?

What value do we create?

What impact do we want to have?

Organizations with clear purpose often navigate uncertainty more successfully because people understand what truly matters.

Purpose becomes a compass.

Technology becomes a vehicle.

The vehicle may change.

The destination remains clear.

Trust Accelerates Performance

Throughout this book, trust has appeared repeatedly.

For good reason.

Trust influences nearly every aspect of organizational performance.

Communication improves.

Collaboration increases.

Innovation accelerates.

Decision-making speeds up.

Engagement rises.

Customer loyalty strengthens.

Trust reduces friction.

When trust is high, organizations move faster.

When trust is low, everything becomes harder.

The future will reward organizations that invest in trust intentionally.

Trust is not a soft concept.

It is a strategic asset.

The Power of Human Judgment

Artificial intelligence can analyze information.

Generate recommendations.

Identify patterns.

Support decisions.

Human beings contribute something equally important:

Judgment.

Judgment evaluates context.

Judgment considers consequences.

Judgment balances competing priorities.

Judgment reflects values.

Judgment applies wisdom.

Organizations that rely exclusively on technology risk making technically correct decisions that create human problems.

The strongest organizations combine analytical capability with human judgment.

This combination produces better outcomes.

Relationships Still Matter

Business has always been built on relationships.

Customer relationships.

Employee relationships.

Partner relationships.

Investor relationships.

Community relationships.

Artificial intelligence can support these relationships.

It cannot replace them.

Trust is built through interaction.

Respect.

Listening.

Empathy.

Shared experiences.

These remain deeply human capabilities.

Organizations that preserve and strengthen relationships will continue creating value long after specific technologies evolve.

The Human Advantage Framework

As leaders prepare for the future, they should focus on strengthening five enduring advantages.

Leadership

Creating clarity, alignment, and direction.

Culture

Building environments where people thrive.

Trust

Establishing credibility and confidence.

Learning

Developing adaptability and capability.

Purpose

Providing meaning and direction.

Together, these elements form what I call the Human Advantage.

Technology amplifies them.

It does not replace them.

The Organization of the Future

The highest-performing organizations of the future will not be defined solely by their technology.

They will be defined by their ability to integrate:

Human intelligence.

Artificial intelligence.

Purpose.

Leadership.

Culture.

Learning.

Trust.

Innovation.

These organizations will not view technology and humanity as competing forces.

They will view them as complementary strengths.

This mindset creates extraordinary possibilities.

A Final Leadership Reflection

Imagine your organization five years from now.

Technology has advanced.

Artificial intelligence is more powerful.

Customers expect more.

Competition is intense.

What will determine success?

Will it be software?

Algorithms?

Platforms?

Partly.

But the larger determinants will remain:

The quality of leadership.

The strength of culture.

The level of trust.

The commitment to learning.

The clarity of purpose.

The development of people.

These factors have always mattered.

They will continue to matter.

The Call to Action

The age of AI is not something to fear.

Nor is it something to worship.

It is something to lead.

The leaders who succeed will embrace technology while strengthening humanity.

They will pursue innovation while preserving values.

They will accelerate performance while developing people.

They will use artificial intelligence to enhance human potential rather than diminish it.

Most importantly, they will recognize that the future belongs not to organizations that choose between people and technology.

The future belongs to organizations that combine the best of both.

Key Takeaways

1. Organizations remain fundamentally human enterprises.
2. Technology creates opportunity, but leadership creates results.
3. Culture remains a primary driver of organizational performance.
4. Purpose provides stability during periods of change.
5. Trust accelerates collaboration, innovation, and execution.
6. Human judgment remains essential.

7. Relationships continue to create value.
8. The Human Advantage will remain the ultimate competitive advantage.

The future will reward organizations that learn faster, adapt faster, and create value faster.

But above all, it will reward organizations that never lose sight of the people they serve and the people who make success possible.

Artificial intelligence may transform how organizations operate.

Leadership, purpose, culture, trust, and human potential will determine how far they go.

IN THE
AGE OF AI

CONCLUSION

The Future Belongs to Learning Organizations

The age of AI has arrived.

The question is no longer whether artificial intelligence will influence your organization.

The question is how intentionally you will lead that influence.

The organizations that thrive will not be those that simply adopt technology.

They will be those that redesign themselves for a new era.

They will build cultures of learning.

Develop adaptable people.

Create trust.

Strengthen leadership.

Align technology with purpose.

And continuously seek new ways to create value.

The future belongs to learning organizations.

Organizations that remain curious.

Organizations that remain adaptable.

Organizations that remain human.

Your journey does not begin with technology.

It begins with leadership.

It begins with purpose.

It begins with a decision.

A decision to learn.

A decision to adapt.

A decision to lead.

The future is already arriving.

The opportunity is yours to create.

Appendix A

The AI Readiness Assessment

How Prepared Is Your Organization for the Age of AI?

Many leaders ask the wrong question.

They ask:

"Are we ready for AI?"

The better question is:

"How ready is our organization to successfully leverage AI?"

Successful AI implementation is not determined by technology alone.

It depends upon leadership, culture, strategy, data, workforce capability, decision-making processes, customer focus, and organizational adaptability.

This assessment is designed to help leaders evaluate their organization's readiness across the areas that matter most.

How to Use This Assessment

For each statement, rate your organization using the following scale:

1 = Strongly Disagree

2 = Disagree

3 = Neutral / Unsure

4 = Agree

5 = Strongly Agree

Record your score for each question and total the results at the end.

SECTION 1

Leadership & Vision

1.

Our leadership team has a clear understanding of how AI could impact our industry.

Score: _____

2.

Our leadership team regularly discusses AI opportunities and risks.

Score: _____

3.

We have a clear vision for how AI supports our business strategy.

Score: _____

4.

Leaders actively encourage innovation and experimentation.

Score: _____

5.

Leadership communicates a compelling future vision for the organization.

Score: _____

Section Total: _____

SECTION 2

Organizational Culture

6.

Employees are encouraged to learn new skills.

Score: _____

7.

People feel safe asking questions and challenging assumptions.

Score: _____

8.

Innovation is encouraged throughout the organization.

Score: _____

9.

Employees generally embrace change.

Score: _____

10.

Continuous improvement is part of our culture.

Score: _____

Section Total: _____

SECTION 3

Workforce Capability

11.

Employees are comfortable using new technology.

Score: _____

12.

We provide ongoing development opportunities.

Score: _____

13.

Our workforce demonstrates strong adaptability.

Score: _____

14.

Employees possess strong critical-thinking skills.

Score: _____

15.

We are actively developing future-ready talent.

Score: _____

Section Total: _____

SECTION 4

Data Readiness

16.

Our data is accurate and reliable.

Score: _____

17.

Important information is easily accessible.

Score: _____

18.

Departments effectively share information.

Score: _____

19.

We use data regularly in decision-making.

Score: _____

20.

We have clear data governance practices.

Score: _____

Section Total: _____

SECTION 5

Decision-Making

21.

Important decisions are made in a timely manner.

Score: _____

22.

We use evidence to support decisions.

Score: _____

23.

Employees have access to information needed to make decisions.

Score: _____

24.

Decision-making authority is clear.

Score: _____

25.

Leaders encourage constructive debate and discussion.

Score: _____

Section Total: _____

SECTION 6

Customer Focus

26.

We deeply understand customer needs.

Score: _____

27.

Customer feedback is regularly collected and reviewed.

Score: _____

28.

Customer experience is a strategic priority.

Score: _____

29.

We respond quickly to customer concerns.

Score: _____

30.

We actively seek opportunities to create more customer value.

Score: _____

Section Total: _____

SECTION 7

Innovation & Adaptability

31.

Employees are encouraged to experiment.

Score: _____

32.

We regularly evaluate new opportunities.

Score: _____

33.

The organization adapts quickly to change.

Score: _____

34.

Cross-functional collaboration is strong.

Score: _____

35.

We continuously seek ways to improve.

Score: _____

Section Total: _____

SECTION 8

Trust & Communication

36.

Employees trust organizational leadership.

Score: _____

37.

Communication is open and transparent.

Score: _____

38.

People understand organizational priorities.

Score: _____

39.

Leaders conduct information checks to verify understanding.

Score: _____

40.

There is a high level of accountability throughout the organization.

Score: _____

Section Total: _____

SECTION 9

AI Readiness

41.

We have identified potential AI use cases.

Score: _____

42.

Employees understand basic AI concepts.

Score: _____

43.

We are actively exploring AI opportunities.

Score: _____

44.

Leadership understands AI-related risks.

Score: _____

45.

We have begun developing an AI strategy.

Score: _____

Section Total: _____

SECTION 10

Future Readiness

46.

Our organization learns quickly.

Score: _____

47.

We are preparing employees for future skill requirements.

Score: _____

48.

We continuously evaluate emerging trends.

Score: _____

49.

Our organizational design supports adaptability.

Score: _____

50.

We are confident in our ability to thrive over the next decade.

Score: _____

Section Total: _____

SCORING

Total Possible Score: 250

Add all section totals:

Total Score: _____

220–250

AI-Ready Leader

Your organization demonstrates many characteristics of a high-performance, future-ready organization. Focus on accelerating execution and scaling AI-enabled innovation.\

180–219

Strong Foundation

Your organization has many strengths but should address gaps before pursuing large-scale AI transformation.

130–179

Developing Capability

You have begun the journey, but significant opportunities exist in leadership, culture, workforce readiness, and organizational systems.

Below 130

Transformation Needed

Your organization should focus first on strengthening leadership, culture, trust, workforce development, and operational systems before making substantial AI investments.

Reflection Questions

1. Which section scored highest?
2. Which section scored lowest?
3. What three improvements would most strengthen AI readiness?
4. What barriers could prevent progress?
5. What actions should be taken in the next 90 days?
6. Who should lead the effort?
7. How will success be measured?
8. What resources will be required?

Remember:

Technology readiness matters.

Organizational readiness matters more.

The organizations that thrive in the AI era are not necessarily the ones with the best technology.

They are the ones with the strongest leadership, healthiest culture, most adaptable workforce, and clearest sense of purpose.

Next: Appendix B , Twenty Questions Every Leadership Team Should Ask About AI. This appendix offers an excellent boardroom and executive team discussion guide.

APPENDIX B

Twenty Questions Every Leadership Team Should Ask About AI

A Boardroom and Executive Team Discussion Guide

One of the greatest risks facing organizations today is not moving too slowly.

It is moving without asking the right questions.

Artificial intelligence creates enormous opportunities.

It also introduces important decisions.

Leaders must determine:

Where to invest.

What to prioritize.

How to prepare the workforce.

How to manage risk.

How to create value.

How to lead responsibly.

The organizations that benefit most from AI are often not those with the most technology.

They are those with the best conversations.

The purpose of this appendix is to help leadership teams, executive committees, boards of directors, and management teams engage in meaningful dialogue about the future.

Use these questions as discussion starters during strategic planning sessions, leadership retreats, board meetings, and executive team discussions.

The goal is not finding perfect answers.

The goal is asking better questions.

SECTION 1

Strategy & Competitive Advantage

1.

How is AI likely to change our industry over the next three to five years?

2.

If a startup were built today using AI from the ground up, how would it compete against us?

3.

What opportunities does AI create to improve our value proposition?

4.

What competitive risks do we face if we delay AI adoption?

5.

How can AI help us achieve our strategic objectives faster or more effectively?

SECTION 2

Customers & Value Creation

6.

How can AI help us better understand our customers?

7.

Where do customers experience the most friction when interacting with us?

8.

How could AI improve customer experience, responsiveness, and personalization?

9.

What new products or services might AI enable us to offer?

10.

How can we use AI to create more value rather than simply reduce costs?

SECTION 3

Workforce & Talent

11.

What skills will become more important in our organization over the next five years?

12.

How prepared is our workforce to work alongside AI?

13.

What development programs are needed to improve AI literacy and adaptability?

14.

How can AI help employees become more productive, creative, and effective?

15.

What concerns might employees have about AI adoption, and how will we address them?

SECTION 4

Leadership & Organizational Design

16.

How must leadership evolve to succeed in an AI-enabled organization?

17.

What organizational structures, processes, or decision-making systems need to change?

18.

How can we improve the speed and quality of decisions using AI?

SECTION 5

Trust, Ethics & Governance

19.

What principles should guide our use of AI?

20.

How will we ensure transparency, accountability, trust, and responsible use of AI?

Bonus Questions for Exceptional Leadership Teams

The strongest leadership teams often go beyond operational discussions and explore deeper strategic questions.

Consider adding these questions to your next retreat or planning session.

Bonus Question 1

What opportunities are becoming possible today that were impossible five years ago?

Bonus Question 2

What assumptions about our business may no longer be true?

Bonus Question 3

If we were starting this organization from scratch today, how would we design it differently?

Bonus Question 4

What capabilities will create the greatest value for customers in the future?

Bonus Question 5

How do we ensure technology strengthens our culture rather than weakens it?

The Executive Team Exercise

Conduct a leadership meeting using the following process:

Step 1

Select five questions from this appendix.

Step 2

Give each participant five minutes to write individual responses.

Step 3

Share responses in pairs.

Step 4

Discuss insights as a full team.

Step 5

Identify common themes.

Step 6

Determine the three most important actions.

Step 7

Assign ownership and timelines.

This process often generates more insight than traditional brainstorming because it combines reflection, dialogue, and accountability.

Board Discussion Framework

Boards should periodically evaluate AI through five lenses:

Opportunity

How can AI create value?

Risk

What could go wrong?

Capability

Do we possess the necessary talent and resources?

Governance

Are safeguards and accountability in place?

Impact

How will AI influence customers, employees, shareholders, and stakeholders?

Boards that consistently evaluate these dimensions provide stronger oversight and support better strategic decisions.

The Most Important Question

Of all the questions in this appendix, one may be more important than the rest:

How can we use AI to help our organization better fulfill its mission?

This question keeps the focus where it belongs.

Not on technology.

Not on trends.

Not on competitors.

On purpose.

Because organizations that align AI with mission create more sustainable value than organizations that pursue technology for its own sake.

Leadership Reflection

As you review these questions, ask yourself:

- Which questions generate the most discussion?
- Which questions make people uncomfortable?
- Which questions have we never asked?
- Which questions reveal our greatest opportunities?
- Which questions reveal our greatest risks?

The quality of an organization's future is often determined by the quality of the questions its leaders are willing to ask.

The leaders who ask better questions see opportunities others miss.

And the organizations that ask better questions often create better futures.

The next appendix translates those questions into action by providing a practical 90-Day AI Transformation Roadmap that leaders can immediately begin implementing within their organizations.

Next: Appendix C , The 90-Day AI Transformation Roadmap. This final appendix provides a concrete step-by-step implementation plan and makes the guidance provided within the pages of this book highly actionable for CEOs, executives, consultants, and business owners.

IN THE
AGE OF AI

APPENDIX C

The 90-Day AI Transformation Roadmap

A Practical Guide for Leaders

One of the most common mistakes organizations make with artificial intelligence is trying to do too much, too quickly.

They purchase multiple tools.

Launch numerous initiatives.

Create committees.

Schedule meetings.

Develop reports.

And six months later, little has changed.

The opposite approach is equally problematic.

Organizations spend months discussing AI without taking meaningful action.

They study.

Research.

Debate.

Analyze.

And never begin.

High-performance organizations avoid both extremes.

They move deliberately.

They focus on outcomes.

They start small.

They learn quickly.

They scale intelligently.

This roadmap is designed to help leaders take meaningful action over the next 90 days.

The goal is not to become an AI-powered organization overnight.

The goal is to establish momentum, build capability, create confidence, and lay the foundation for long-term success.

Phase One

Days 1–30

Assess, Align, and Educate

Objective

Create organizational awareness, leadership alignment, and a clear understanding of opportunities.

Step 1

Complete the AI Readiness Assessment

Use Appendix A to evaluate your organization's current state.

Focus on:

- Leadership readiness
- Workforce readiness
- Data readiness
- Cultural readiness
- Organizational readiness

Deliverable

Current State Assessment Report

Step 2

Conduct an Executive Leadership Session

Bring together key leaders.

Discuss:

- Opportunities
- Risks
- Industry implications
- Organizational priorities

Use Appendix B as the discussion guide.

Deliverable

Leadership Alignment Summary

Step 3

Define Success

Answer three questions:

What are we trying to achieve?

How will success be measured?

What business outcomes matter most?

Examples:

- Productivity improvement
- Faster decision-making

- Better customer experience
- Revenue growth
- Cost reduction

Deliverable

AI Success Statement

Step 4

Identify High-Value Use Cases

Select three to five opportunities.

Look for:

- Repetitive work
- Information-intensive processes
- Customer experience improvements
- Decision-support opportunities

Avoid trying to transform everything at once.

Deliverable

Priority Opportunity List

Step 5

Begin Leadership AI Training

Every executive should develop basic AI literacy.

Topics include:

- AI fundamentals
- Opportunities

- Risks
- Governance
- Practical applications

Deliverable

Leadership Learning Plan

Days 1–30 Success Indicators

✓ Leadership alignment established

✓ Readiness assessment completed

✓ Success measures defined

✓ Initial opportunities identified

✓ Leadership training underway

Phase Two

Days 31–60

Experiment and Build Capability

Objective

Create momentum through practical experimentation.

Step 6

Launch Pilot Projects

Choose one to three pilot initiatives.

Examples:

- Meeting summaries

- Content creation
- Customer support enhancement
- Knowledge management
- Reporting automation

Keep pilots manageable.

Focus on learning.

Deliverable

Pilot Project Charter

Step 7

Create an AI Champion Team

Identify employees who are:

- Curious
- Adaptable
- Influential
- Open to experimentation

These individuals become internal advocates.

Deliverable

AI Champion Network

Step 8

Begin Workforce Education

Introduce employees to:

- AI fundamentals

- Responsible use
- Productivity applications
- Human-AI collaboration

Focus on confidence rather than complexity.

Deliverable

Workforce Learning Program

Step 9

Establish Governance Principles

Create simple guidelines addressing:

- Transparency
- Data privacy
- Security
- Accountability
- Ethical use

Deliverable

AI Guiding Principles Document

Step 10

Capture Lessons Learned

Document:

- Successes
- Challenges
- Employee feedback
- Customer feedback

- Improvement opportunities

Deliverable

Pilot Learning Report

Days 31–60 Success Indicators

✓ Pilot projects active

✓ Employee engagement increasing

✓ AI champions identified

✓ Governance principles established

✓ Organizational learning accelerating

Phase Three

Days 61–90

Scale and Integrate

Objective

Expand successful initiatives and integrate AI into daily operations.

Step 11

Evaluate Pilot Results

Assess:

- Productivity gains
- Quality improvements
- Employee adoption
- Customer impact

- Lessons learned

Deliverable

Pilot Evaluation Report

Step 12

Expand Successful Use Cases

Scale initiatives that produced measurable value.

Avoid scaling projects that lack evidence of success.

Deliverable

Expansion Plan

Step 13

Integrate AI into Workflows

Move beyond experimentation.

Embed successful practices into:

- Daily work
- Team processes
- Decision-making
- Communication
- Customer interactions

Deliverable

Updated Operating Procedures

Step 14

Develop an AI Strategy

Create a 12–24 month roadmap addressing:

- Technology
- Workforce development
- Data strategy
- Customer experience
- Innovation
- Growth

Deliverable

AI Strategic Roadmap

Step 15

Communicate Success Stories

Share examples of:

- Improved performance
- Employee growth
- Customer value
- Innovation outcomes

Stories build momentum.

Momentum accelerates adoption.

Deliverable

Transformation Success Report

Days 61–90 Success Indicators

✓ Successful pilots scaled

✓ AI integrated into workflows

✓ Strategic roadmap developed

✓ Employee confidence increasing

✓ Measurable business value demonstrated

The Five Rules of Successful AI Transformation

Throughout the first 90 days, remember these principles.

Rule 1

Start with Purpose

Focus on business outcomes, not technology.

Rule 2

Think Small to Grow Big

Begin with manageable wins.

Build momentum.

Rule 3

Develop People First

Technology adoption follows capability development.

Rule 4

Measure Outcomes

Track value, not activity.

Rule 5

Learn Continuously

Treat every initiative as a learning opportunity.

Common Mistakes to Avoid

Mistake #1

Starting with technology rather than business objectives.

Mistake #2

Trying to transform everything at once.

Mistake #3

Ignoring employee concerns.

Mistake #4

Failing to establish governance.

Mistake #5

Measuring activity rather than outcomes.

Mistake #6

Assuming implementation equals transformation.

Mistake #7

Treating AI as an IT project rather than an organizational transformation initiative.

The Leadership Dashboard

Track the following monthly:

Leadership

- Leadership AI literacy
- Executive engagement

Workforce

- Employee participation
- Training completion
- AI confidence levels

Operations

- Productivity improvements
- Cycle-time reductions
- Quality improvements

Customers

- Satisfaction scores
- Response times
- Retention rates

Innovation

- New ideas generated
- Pilot projects launched
- Opportunities identified

Growth

- Revenue opportunities
- Market expansion initiatives
- New services developed

The Final Thought

The goal of AI transformation is not implementing artificial intelligence.

The goal is creating a stronger organization.

A more adaptable organization.

A more innovative organization.

A more productive organization.

A more customer-focused organization.

A more human-centered organization.

Artificial intelligence is a powerful tool.

But tools do not create transformation.

Leaders do.

People do.

Organizations do.

The future belongs to leaders who combine technology, purpose, culture, trust, learning, and execution into a unified system for creating value.

Your first 90 days begin now.

ABOUT THE AUTHOR

Wilts C. Alexander III is an executive coach, leadership consultant, author, speaker, and organizational development strategist.

For more than five decades, he has worked with executives, entrepreneurs, nonprofit leaders, educators, and organizations seeking to improve performance, develop leaders, create stronger cultures, and achieve sustainable results.

His work focuses on the intersection of leadership, organizational effectiveness, personal development, and emerging technologies, helping leaders navigate change while maintaining a strong commitment to people, purpose, and performance.

Wilts believes that the future belongs to learning organizations, organizations that continuously develop people, embrace innovation, strengthen culture, and create meaningful value for the communities they serve.

He lives by a simple principle:

Technology should enhance human potential, not replace it.

ACKNOWLEDGMENTS

No book is ever written alone.

While my name appears on the cover, the ideas contained in these pages have been shaped by thousands of conversations with executives, entrepreneurs, managers, coaches, educators, nonprofit leaders, and lifelong learners.

I am grateful to the leaders who opened their organizations and allowed me to observe what creates extraordinary performance and what causes organizations to struggle.

I am grateful to the clients who challenged my thinking, asked difficult questions, and trusted me to support their growth.

I am grateful to the pioneers in artificial intelligence, leadership, organizational development, and human performance whose work continues to influence how we think about the future.

Most importantly, I am grateful to the individuals who believe that leadership is not about power but about service, stewardship, and helping others reach their potential.

This book is for you.

Final Reflection

The age of AI will reward organizations that learn faster, adapt faster, and create value faster.

But above all, it will reward organizations that never forget a simple truth:

Technology may power the future.

People create it.

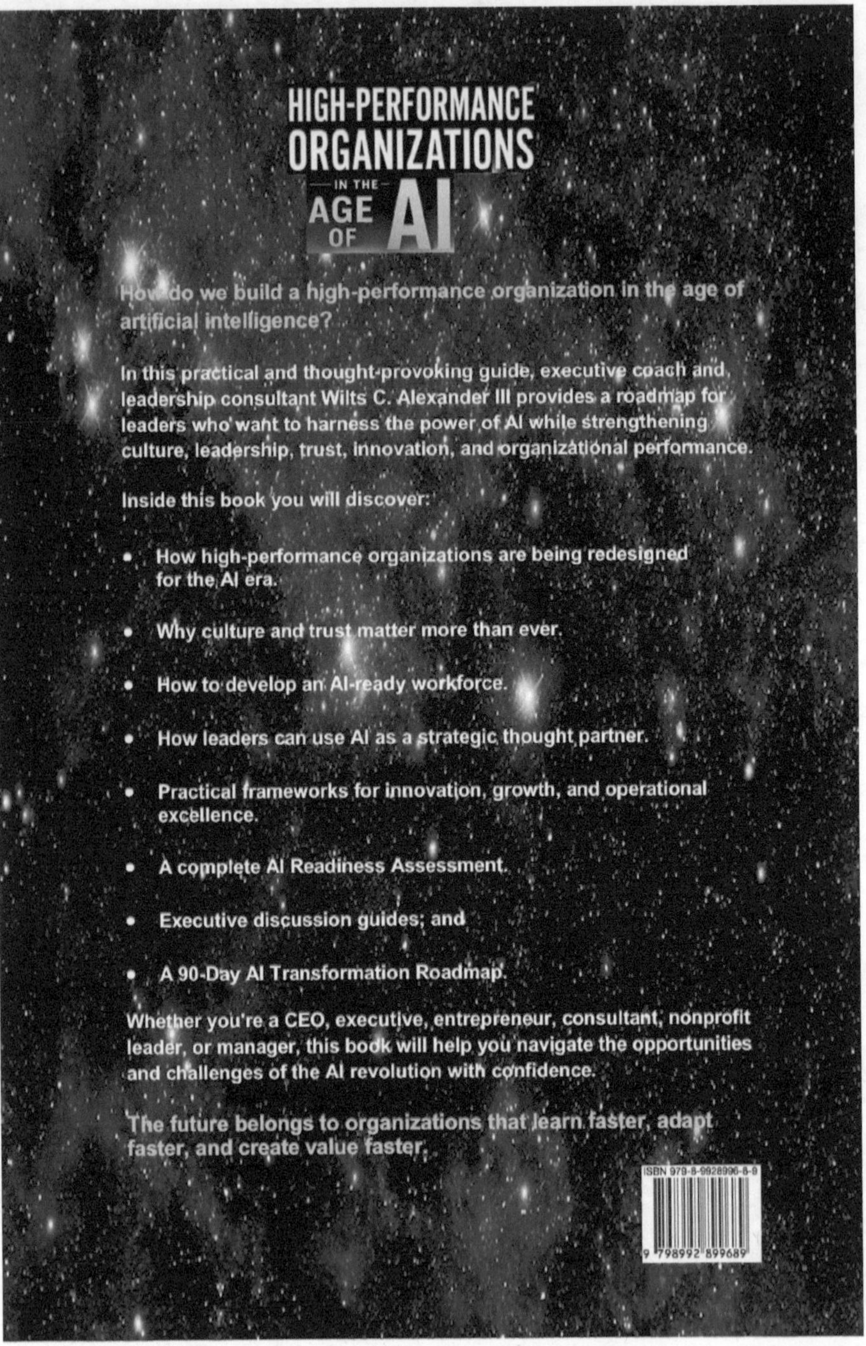

HIGH-PERFORMANCE ORGANIZATIONS IN THE AGE OF AI

How do we build a high-performance organization in the age of artificial intelligence?

In this practical and thought-provoking guide, executive coach and leadership consultant Wilts C. Alexander III provides a roadmap for leaders who want to harness the power of AI while strengthening culture, leadership, trust, innovation, and organizational performance.

Inside this book you will discover:

- How high-performance organizations are being redesigned for the AI era.
- Why culture and trust matter more than ever.
- How to develop an AI-ready workforce.
- How leaders can use AI as a strategic thought partner.
- Practical frameworks for innovation, growth, and operational excellence.
- A complete AI Readiness Assessment.
- Executive discussion guides; and
- A 90-Day AI Transformation Roadmap.

Whether you're a CEO, executive, entrepreneur, consultant, nonprofit leader, or manager, this book will help you navigate the opportunities and challenges of the AI revolution with confidence.

The future belongs to organizations that learn faster, adapt faster, and create value faster.

ISBN 979-8-9928996-8-9

9 798992 899689

www.ingramcontent.com/pod-product-compliance
Lightning Source LLC
LaVergne TN
LVHW100135061126
851857LV00010B/103